Deep Learning By Example

A hands-on guide to implementing advanced machine
learning algorithms and neural networks

Ahmed Menshawy

BIRMINGHAM - MUMBAI

Deep Learning By Example

Commissioning Editor: Vedika Naik
Acquisition Editor: Tushar Gupta
Content Development Editor: Aishwarya Pandere
Technical Editor: Sagar Sawant
Copy Editor: Vikrant Phadke, Safis Editing
Project Coordinator: Nidhi Joshi
Proofreader: Safis Editing
Indexer: Mariammal Chettiyar
Graphics: Tania Dutta
Production Coordinator: Aparna Bhagat

First published: February 2018

Production reference: 1260218

Published by Packt Publishing Ltd.
Livery Place
35 Livery Street
Birmingham
B3 2PB, UK.

ISBN 978-1-78839-990-6

www.packtpub.com

`mapt.io`

Mapt is an online digital library that gives you full access to over 5,000 books and videos, as well as industry leading tools to help you plan your personal development and advance your career. For more information, please visit our website.

Why subscribe?

- Spend less time learning and more time coding with practical eBooks and Videos from over 4,000 industry professionals

- Improve your learning with Skill Plans built especially for you

- Get a free eBook or video every month

- Mapt is fully searchable

- Copy and paste, print, and bookmark content

PacktPub.com

Did you know that Packt offers eBook versions of every book published, with PDF and ePub files available? You can upgrade to the eBook version at `www.PacktPub.com` and as a print book customer, you are entitled to a discount on the eBook copy. Get in touch with us at `service@packtpub.com` for more details.

At `www.PacktPub.com`, you can also read a collection of free technical articles, sign up for a range of free newsletters, and receive exclusive discounts and offers on Packt books and eBooks.

Contributors

About the author

Ahmed Menshawy is a Research Engineer at the Trinity College Dublin, Ireland. He has more than 5 years of working experience in the area of ML and NLP. He holds an MSc in Advanced Computer Science. He started his Career as a Teaching Assistant at the Department of Computer Science, Helwan University, Cairo, Egypt. He taught several advanced ML and NLP courses such as ML, Image Processing, and so on. He was involved in implementing the state-of-the-art system for Arabic Text to Speech. He was the main ML specialist at the Industrial research and development lab at IST Networks, based in Egypt.

I want to thank the people who have been close to me and supported me, especially my wife Sara and my parents

About the reviewers

Md. Rezaul Karim is a Research Scientist at Fraunhofer FIT, Germany. He is also a PhD candidate at RWTH Aachen University, Germany. Before joining FIT, he worked as a Researcher at the Insight Centre for Data Analytics, Ireland. Earlier, he worked as a Lead Engineer at Samsung Electronics, Korea.

He has 9 years of R&D experience with C++, Java, R, Scala, and Python. He has published research papers concerning bioinformatics, big data, and deep learning. He has practical working experience with Spark, Zeppelin, Hadoop, Keras, Scikit-Learn, TensorFlow, DeepLearning4j, MXNet, and H2O.

Doug Ortiz is an experienced Enterprise Cloud, Big Data, Data Analytics and Solutions Architect who has architected, designed, developed, re-engineered and integrated enterprise solutions. Other expertise: Amazon Web Services, Azure, Google Cloud, Business Intelligence, Hadoop, Spark, NoSQL Databases, SharePoint to mention a few.

Is the founder of Illustris, LLC reachable at: dougortiz@illustris.org

Huge thanks to my wonderful wife Milla, Maria, Nikolay and our children for all their support.

Packt is searching for authors like you

If you're interested in becoming an author for Packt, please visit authors.packtpub.com and apply today. We have worked with thousands of developers and tech professionals, just like you, to help them share their insight with the global tech community. You can make a general application, apply for a specific hot topic that we are recruiting an author for, or submit your own idea.

Table of Contents

Preface

This book will start off by introducing the foundations of machine learning, what makes learning visible, demonstrating the traditional machine learning techniques with some examples and eventually deep learning. You will then move to creating machine learning models that will eventually lead you to neural networks. You will get familiar with the basics of deep learning and explore various tools that enable deep learning in a powerful yet user-friendly manner. With a very low starting point, this book will enable a regular developer to get hands-on experience with deep learning. You will learn all the essentials needed to explore and understand what deep learning is and will perform deep learning tasks first-hand. Also, we will be using one of the most widely used deep learning frameworks. TensorFlow has big community support that is growing day by day, which makes it a good option for building your complex deep learning applications.

Who this book is for

This book is a starting point for those who are keen on knowing about deep learning and implementing it but do not have an extensive background in machine learning, complex statistics, and linear algebra.

What this book covers

Chapter 1, *Data science - Bird's-eye view*, explains that data science or machine learning is the process of giving the machines the ability to learn from a dataset without being told or programmed. For instance, it will be extremely hard to write a program that takes a hand-written digit as an input image and outputs a value from 0-9 according to the number that's written in this image. The same applies to the task of classifying incoming emails as spam or non-spam. To solve such tasks, data scientists uses learning methods and tools from the field of data science or machine learning to teach the computer how to automatically recognize digits by giving it some explanatory features that can distinguish each digit from another. The same for the spam/non-spam problem, instead of using regular expressions and writing hundred of rules to classify the incoming emails, we can teach the computer through specific learning algorithms how to distinguish between spam and non-spam emails.

Chapter 2, *Getting Started with Data Science – Titanic Example,* linear models are the basic learning algorithms in the field of data science. Understanding how a linear model works is crucial in your journey of learning data science because it's the basic building block for most of the sophisticated learning algorithms out there, including neural networks.

Chapter 3, *Feature Engineering and Model Complexity – Titanic Example Revisited,* covers model complexity and assessment. This is an important towards building a successful data science system. There are lots of tools that you can use to assess and choose your model. In this chapter, we are going to address some of tools that can help you to increase the value of your data by adding more descriptive features and extracting meaningful information from existing ones. We are also going to address other tools related to optimal number features and learn why it's a problem to have a large number of features and fewer training samples/observations.

Chapter 4, *Get Up and Running with TensorFlow,* gives an overview of one of the most widely used deep learning frameworks. TensorFlow has big community support that is growing day by day, which makes it a good option for building your complex deep learning applications

Chapter 5, *Tensorflow in Action - Some Basic Examples,* will explain the main computational concept behind TensorFlow, which is the computational graph model, and demonstrate how to get you on track by implementing linear regression and logistic regression.

Chapter 6, *Deep Feed-forward Neural Networks - Implementing Digit Classification,* explains that a **feed-forward neural network** (**FNN**) is a special type of neural network wherein links/connections between neurons do not form a cycle. As such, it is different from other architectures in a neural network that we will get to study later on in this book (recurrent-type neural networks). The FNN is a widely used architecture and it was the first and simplest type of neural network. In this chapter, we will go through the architecture of a typical ;FNN, and we will be using the TensorFlow library for this. After covering these concepts, we will give a practical example of digit classification. The question of this example is, *Given a set of images that contain handwritten digits, how can you classify these images into 10 different classes (0-9)?*

Chapter 7, *Introduction to Convolutional Neural Networks*, explains that in data science, a **convolutional neural network (CNN)** is specific kind of deep learning architecture that uses the convolution operation to extract relevant explanatory features for the input image. CNN layers are connected as an FNN while using this convolution operation to mimic how the human brain functions when trying to recognize objects. Individual cortical neurons respond to stimuli in a restricted region of space known as the receptive field. In particular, biomedical imaging problems could be challenge sometimes but in this chapter, we'll see how to use a CNN in order to discover patterns in this image.

Chapter 8, *Object Detection – CIFAR-10 Example*, covers the basics and the intuition/motivation behind CNNs, before demonstrating this on one of the most popular datasets available for object detection. We'll also see how the initial layers of the CNN get very basic features about our objects, but the final convolutional layers will get more semantic-level features that are built up from those basic features in the first layers.

Chapter 9, *Object Detection – Transfer Learning with CNNs*, explains that **Transfer learning (TL)** is a research problem in data science that is mainly concerned with persisting knowledge acquired during solving a specific task and using this acquired knowledge to solve another different but similar task. In this chapter, we will demonstrate one of the modern practices and common themes used in the field of data science with TL. The idea here is how to get the help from domains with very large datasets to domains that have smaller datasets. Finally, we will revisit our object detection example of CIFAR-10 and try to reduce both the training time and performance error via TL.

Chapter 10, *Recurrent-Type Neural Networks - Language Modeling*, explains that **Recurrent neural networks (RNNs)** are a class of deep learning architectures that are widely used for natural language processing. This set of architectures enables us to provide contextual information for current predictions and also have specific architecture that deals with long-term dependencies in any input sequence. In this chapter, we'll demonstrate how to make a sequence-to-sequence model, which will be useful in many applications in NLP. We will demonstrate these concepts by building a character-level language model and see how our model generates sentences similar to original input sequences.

Chapter 11, *Representation Learning - Implementing Word Embeddings*, explains that machine learning is a science that is mainly based on statistics and linear algebra. Applying matrix operations is very common among most machine learning or deep learning architectures because of backpropagation. This is the main reason deep learning, or machine learning in general, accepts only real-valued quantities as input. This fact contradicts many applications, such as machine translation, sentiment analysis, and so on; they have text as an input. So, in order to use deep learning for this application, we need to have it in the form that deep learning accepts! In this chapter, we are going to introduce the field of representation learning, which is a way to learn a real-valued representation from text while preserving the semantics of the actual text. For example, the representation of love should be very close to the representation of adore because they are used in very similar contexts.

Chapter 12, *Neural Sentiment Analysis*, addresses one of the hot and trendy applications in natural language processing, which is called sentiment analysis. Most people nowadays express their opinions about something through social media platforms, and making use of this vast amount of text to keep track of customer satisfaction about something is very crucial for companies or even governments.

In this chapter, we are going to use RNNs to build a sentiment analysis solution.

Chapter 13, *Autoencoders – Feature Extraction and Denoising*, explains that an autoencoder network is nowadays one of the widely used deep learning architectures. It's mainly used for unsupervised learning of efficient decoding tasks. It can also be used for dimensionality reduction by learning an encoding or a representation for a specific dataset. Using autoencoders in this chapter, we'll show how to denoise your dataset by constructing another dataset with the same dimensions but less noise. To use this concept in practice, we will extract the important features from the MNIST dataset and try to see how the performance will be significantly enhanced by this.

Chapter 14, *Generative Adversarial Networks*, covers **Generative Adversarial Networks (GANs)**. They are deep neural net architectures that consist of two networks pitted against each other (hence the name adversarial). GANs were introduced in a paper (https://arxiv.org/abs/1406.2661) by Ian Goodfellow and other researchers, including Yoshua Bengio, at the University of Montreal in 2014. Referring to GANs, Facebook's AI research director, Yann LeCun, called **adversarial training** the most interesting idea in the last 10 years in machine learning. The potential of GANs is huge, because they can learn to mimic any distribution of data. That is, GANs can be taught to create worlds eerily similar to our own in any domain: images, music, speech, or prose. They are robot artists in a sense, and their output is impressive (https://www.nytimes.com/2017/08/14/arts/design/google-how-ai-creates-new-music-and-new-artists-project-magenta.html)—and poignant too.

Chapter 15, *Face Generation and Handling Missing Labels*, shows that the list of interesting applications that we can use GANs for is endless. In this chapter, we are going to demonstrate another promising application of GANs, which is face generation based on the CelebA database. We'll also demonstrate how to use GANs for semi-supervised learning setups where we've got a poorly labeled dataset with some missing labels.

Appendix, *Implementing Fish Recognition*, includes entire piece of code of fish recognition example.

To get the most out of this book

- Inform the reader of the things that they need to know before they start, and spell out what knowledge you are assuming.
- Any additional installation instructions and information they need for getting set up.

Download the example code files

You can download the example code files for this book from your account at www.packtpub.com. If you purchased this book elsewhere, you can visit www.packtpub.com/support and register to have the files emailed directly to you.

You can download the code files by following these steps:

1. Log in or register at www.packtpub.com.
2. Select the **SUPPORT** tab.
3. Click on **Code Downloads & Errata**.
4. Enter the name of the book in the **Search** box and follow the onscreen instructions.

Once the file is downloaded, please make sure that you unzip or extract the folder using the latest version of:

- WinRAR/7-Zip for Windows
- Zipeg/iZip/UnRarX for Mac
- 7-Zip/PeaZip for Linux

The code bundle for the book is also hosted on GitHub at
`https://github.com/PacktPublishing/Deep-Learning-By-Example`. We also have other
code bundles from our rich catalog of books and videos available at `https://github.com/`
`PacktPublishing/`. Check them out!

Download the color images

We also provide a PDF file that has color images of the screenshots/diagrams used in this
book. You can download it here: `http://www.packtpub.com/sites/default/files/`
`downloads/`
DeepLearningByExample_ColorImages.pdf.

Conventions used

There are a number of text conventions used throughout this book.

`CodeInText`: Indicates code words in text, database table names, folder names, filenames,
file extensions, pathnames, dummy URLs, user input, and Twitter handles. Here is an
example: "Mount the downloaded `WebStorm-10*.dmg` disk image file as another disk in
your system."

A block of code is set as follows:

```
html, body, #map {
  height: 100%;
  margin: 0;
  padding: 0
}
```

When we wish to draw your attention to a particular part of a code block, the relevant lines
or items are set in bold:

```
[default]
exten => s,1,Dial(Zap/1|30)
exten => s,2,Voicemail(u100)
exten => s,102,Voicemail(b100)
exten => i,1,Voicemail(s0)
```

Any command-line input or output is written as follows:

```
$ mkdir css
$ cd css
```

Bold: Indicates a new term, an important word, or words that you see onscreen. For example, words in menus or dialog boxes appear in the text like this. Here is an example: "Select **System info** from the **Administration** panel."

 Warnings or important notes appear like this.

 Tips and tricks appear like this.

Get in touch

Feedback from our readers is always welcome.

General feedback: Email `feedback@packtpub.com` and mention the book title in the subject of your message. If you have questions about any aspect of this book, please email us at `questions@packtpub.com`.

Errata: Although we have taken every care to ensure the accuracy of our content, mistakes do happen. If you have found a mistake in this book, we would be grateful if you would report this to us. Please visit `www.packtpub.com/submit-errata`, selecting your book, clicking on the Errata Submission Form link, and entering the details.

Piracy: If you come across any illegal copies of our works in any form on the Internet, we would be grateful if you would provide us with the location address or website name. Please contact us at `copyright@packtpub.com` with a link to the material.

If you are interested in becoming an author: If there is a topic that you have expertise in and you are interested in either writing or contributing to a book, please visit `authors.packtpub.com`.

Reviews

Please leave a review. Once you have read and used this book, why not leave a review on the site that you purchased it from? Potential readers can then see and use your unbiased opinion to make purchase decisions, we at Packt can understand what you think about our products, and our authors can see your feedback on their book. Thank you!

For more information about Packt, please visit `packtpub.com`.

1
Data Science - A Birds' Eye View

Data science or machine learning is the process of giving the machines the ability to learn from a dataset without being told or programmed. For instance, it is extremely hard to write a program that can take a hand-written digit as an input image and outputs a value from 0-9 according to the image that's written. The same applies to the task of classifying incoming emails as spam or non-spam. For solving such tasks, data scientists use learning methods and tools from the field of data science or machine learning to teach the computer how to automatically recognize digits, by giving it some explanatory features that can distinguish one digit from another. The same for the spam/non-spam problem, instead of using regular expressions and writing hundred of rules to classify the incoming email, we can teach the computer through specific learning algorithms how to distinguish between spam and non-spam emails.

 For the spam filtering application, you can code it by a rule-based approach, but it won't be good enough to be used in production, like the one in your mailing server. Building a learning system is an ideal solution for that.

You are probably using applications of data science on a daily basis, often without knowing it. For example, your country might be using a system to detect the ZIP code of your posted letter in order to automatically forward it to the correct area. If you are using Amazon, they often recommend things for you to buy and they do this by learning what sort of things you often search for or buy.

Building a learned/trained machine learning algorithm will require a base of historical data samples from which it's going to learn how to distinguish between different examples and to come up with some knowledge and trends from that data. After that, the learned/trained algorithm could be used for making predictions on unseen data. The learning algorithm will be using raw historical data and will try to come up with some knowledge and trends from that data.

In this chapter, we are going to have a bird's-eye view of data science, how it works as a black box, and the challenges that data scientists face on a daily basis. We are going to cover the following topics:

- Understanding data science by an example
- Design procedure of data science algorithms
- Getting to learn
- Implementing the fish recognition/detection model
- Different learning types
- Data size and industry needs

Understanding data science by an example

To illustrate the life cycle and challenges of building a learning algorithm for specific data, let us consider a real example. The Nature Conservancy is working with other fishing companies and partners to monitor fishing activities and preserve fisheries for the future. So they are looking to use cameras in the future to scale up this monitoring process. The amount of data that will be produced from the deployment of these cameras will be cumbersome and very expensive to process manually. So the conservancy wants to develop a learning algorithm to automatically detect and classify different species of fish to speed up the video reviewing process.

Figure 1.1 shows a sample of images taken by conservancy-deployed cameras. These images will be used to build the system.

Figure 1.1: Sample of the conservancy-deployed cameras' output

So our aim in this example is to separate different species such as tunas, sharks, and more that fishing boats catch. As an illustrative example, we can limit the problem to only two classes, tuna and opah.

Figure 1.2: Tuna fish type (left) and opah fish type (right)

After limiting our problem to contain only two types of fish, we can take a sample of some random images from our collection and start to note some physical differences between the two types. For example, consider the following physical differences:

- **Length**: You can see that compared to the opah fish, the tuna fish is longer
- **Width**: Opah is wider than tuna
- **Color**: You can see that the opah fish tends to be more red while the tuna fish tends to be blue and white, and so on

We can use these physical differences as features that can help our learning algorithm(classifier) to differentiate between these two types of fish.

Explanatory features of an object are something that we use in daily life to discriminate between objects that surround us. Even babies use these explanatory features to learn about the surrounding environment. The same for data science, in order to build a learned model that can discriminate between different objects (for example, fish type), we need to give it some explanatory features to learn from (for example, fish length). In order to make the model more certain and reduce the confusion error, we can increase (to some extent) the explanatory features of the objects.

Given that there are physical differences between the two types of fish, these two different fish populations have different models or descriptions. So the ultimate goal of our classification task is to get the classifier to learn these different models and then give an image of one of these two types as an input. The classifier will classify it by choosing the model (tuna model or opah model) that corresponds best to this image.

In this case, the collection of tuna and opah fish will act as the knowledge base for our classifier. Initially, the knowledge base (training samples) will be labeled/tagged, and for each image, you will know beforehand whether it's tuna or opah fish. So the classifier will use these training samples to model the different types of fish, and then we can use the output of the training phase to automatically label unlabeled/untagged fish that the classifier didn't see during the training phase. This kind of unlabeled data is often called **unseen data**. The training phase of the life cycle is shown in the following diagram:

 Supervised data science is all about learning from historical data with known target or output, such as the fish type, and then using this learned model to predict cases or data samples, for which we don't know the target/output.

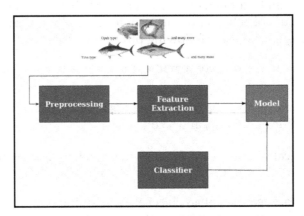

Figure 1.3: Training phase life cycle

Let's have a look at how the training phase of the classifier will work:

- **Pre-processing**: In this step, we will try to segment the fish from the image by using the relevant segmentation technique.
- **Feature extraction**: After segmenting the fish from the image by subtracting the background, we will measure the physical differences (length, width, color, and so on) of each image. At the end, you will get something like *Figure 1.4*.

Finally, we will feed this data into the classifier in order to model different fish types.

As we have seen, we can visually differentiate between tuna and opah fish based on the physical differences (features) that we proposed, such as length, width, and color.

We can use the length feature to differentiate between the two types of fish. So we can try to differentiate between the fish by observing their length and seeing whether it exceeds some value (length*) or not.

So, based on our training sample, we can derive the following rule:

```
If length(fish)> length* then label(fish) = Tuna
Otherwise label(fish) = Opah
```

In order to find this length* we can somehow make length measurements based on our training samples. So, suppose we get these length measurements and obtain the histogram as follows:

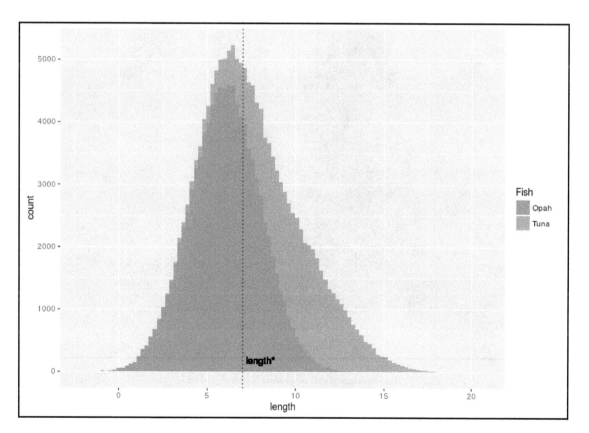

Figure 1.4: Histogram of the length measurements for the two types of fish

In this case, we can derive a rule based on the length feature and differentiate the tuna and opah fish. In this particular example, we can tell that `length*` is 7. So we can update the preceding rule to be:

```
If length(fish)> 7 then label(fish) = Tuna
Otherwise label(fish) = Opah
```

As you may notice, this is not a promising result because of the overlap between the two histograms, as the length feature is not a perfect one to use solely for differentiating between the two types. So we can try to incorporate more features such as the width and then combine them. So, if we somehow manage to measure the width of our training samples, we might get something like the histogram as follows:

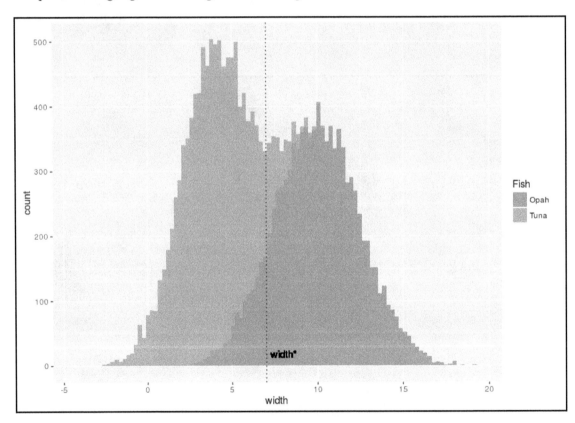

Figure 5: Histogram of width measurements for the two types of fish

As you can see, depending on one feature only will not give accurate results and the output model will do lots of misclassifications. Instead, we can somehow combine the two features and come up with something that looks reasonable.

So if we combine both features, we might get something that looks like the following graph:

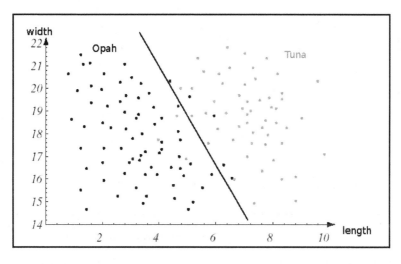

Figure 1.6 : Combination between the subset of the length and width measurements for the two types of fish

Combining the readings for the **length** and **width** features, we will get a scatter plot like the one in the preceding graph. We have the red dots to represent the tuna fish and the green dots to represent the opah fish, and we can suggest this black line to be the rule or the decision boundary that will differentiate between the two types of fish.

For example, if the reading of one fish is above this decision boundary, then it's a tuna fish; otherwise, it will be predicted as an opah fish.

We can somehow try to increase the complexity of the rule to avoid any errors and get a decision boundary like the one in the following graph:

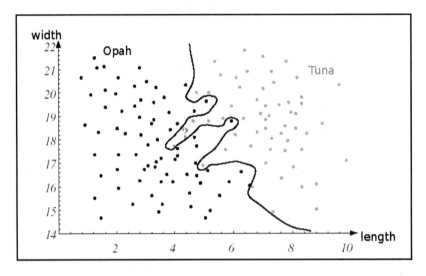

Figure 1.7: Increasing the complexity of the decision boundary to avoid misclassifications over the training data

The advantage of this model is that we get almost 0 misclassifications over the training samples. But actually this is not the objective of using data science. The objective of data science is to build a model that will be able to generalize and perform well over the unseen data. In order to find out whether we built a model that will generalize or not, we are going to introduce a new phase called the **testing phase**, in which we give the trained model an unlabeled image and expect the model to assign the correct label (**Tuna** and **Opah**) to it.

Data science's ultimate objective is to build a model that will work well in production, not over the training set. So don't be happy when you see your model is performing well on the training set, like the one in figure 1.7. Mostly, this kind of model will fail to work well in recognizing the fish type in the image. This incident of having your model work well only over the training set is called **overfitting**, and most practitioners fall into this trap.

Instead of coming up with such a complex model, you can drive a less complex one that will generalize in the testing phase. The following graph shows the use of a less complex model in order to get fewer misclassification errors and to generalize the unseen data as well:

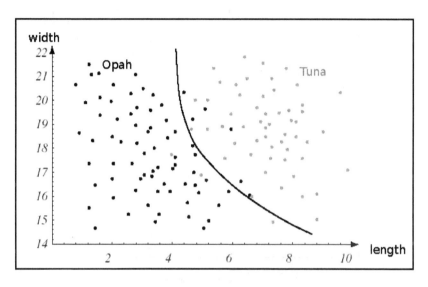

Figure 1.8: Using a less complex model in order to be able to generalize over the testing samples (unseen data)

Design procedure of data science algorithms

Different learning systems usually follow the same design procedure. They start by acquiring the knowledge base, selecting the relevant explanatory features from the data, going through a bunch of candidate learning algorithms while keeping an eye on the performance of each one, and finally the evaluation process, which measures how successful the training process was.

In this section, we are going to address all these different design steps in more detail:

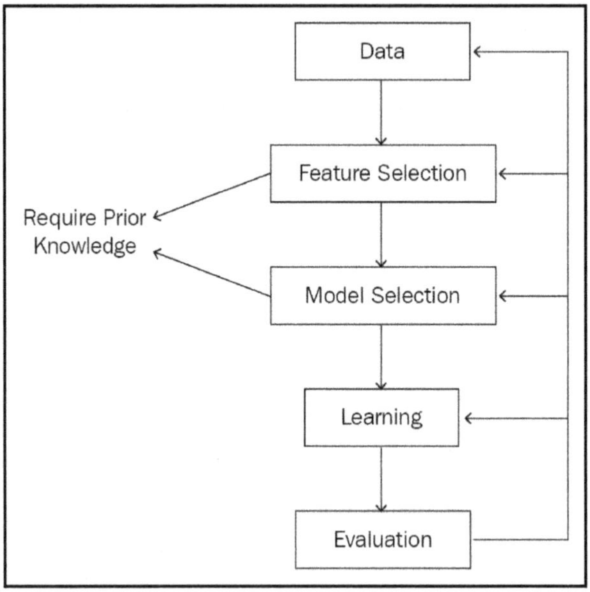

Figure 1.11: Model learning process outline

Data pre-processing

This component of the learning cycle represents the knowledge base of our algorithm. So, in order to help the learning algorithm give accurate decisions about the unseen data, we need to provide this knowledge base in the best form. Thus, our data may need a lot of cleaning and pre-processing (conversions).

Data cleaning

Most datasets require this step, in which you get rid of errors, noise, and redundancies. We need our data to be accurate, complete, reliable, and unbiased, as there are lots of problems that may arise from using bad knowledge base, such as:

- Inaccurate and biased conclusions
- Increased error
- Reduced generalizability, which is the model's ability to perform well over the unseen data that it didn't train on previously

Data pre-processing

In this step, we apply some conversions to our data to make it consistent and concrete. There are lots of different conversions that you can consider while pre-processing your data:

- **Renaming** (**relabeling**): This means converting categorical values to numbers, as categorical values are dangerous if used with some learning methods, and also numbers will impose an order between the values
- **Rescaling** (**normalization**): Transforming/bounding continuous values to some range, typically *[-1, 1]* or *[0, 1]*
- **New features**: Making up new features from the existing ones. For example, *obesity-factor = weight/height*

Feature selection

The number of explanatory features (input variables) of a sample can be enormous wherein you get $x_i=(x_i^1, x_i^2, x_i^3, ... , x_i^d)$ as a training sample (observation/example) and d is very large. An example of this can be a document classification task3, where you get 10,000 different words and the input variables will be the number of occurrences of different words.

This enormous number of input variables can be problematic and sometimes a curse because we have many input variables and few training samples to help us in the learning procedure. To avoid this curse of having an enormous number of input variables (curse of dimensionality), data scientists use dimensionality reduction techniques in order to select a subset from the input variables. For example, in the text classification task they can do the following:

- Extracting relevant inputs (for instance, mutual information measure)
- **Principal component analysis** (**PCA**)
- Grouping (cluster) similar words (this uses a similarity measure)

Model selection

This step comes after selecting a proper subset of your input variables by using any dimensionality reduction technique. Choosing the proper subset of the input variable will make the rest of the learning process very simple.

In this step, you are trying to figure out the right model to learn.

If you have any prior experience with data science and applying learning methods to different domains and different kinds of data, then you will find this step easy as it requires prior knowledge of how your data looks and what assumptions could fit the nature of your data, and based on this you choose the proper learning method. If you don't have any prior knowledge, that's also fine because you can do this step by guessing and trying different learning methods with different parameter settings and choose the one that gives you better performance over the test set.

Also, initial data analysis and visualization will help you to make a good guess about the form of the distribution and nature of your data.

Learning process

By learning, we mean the optimization criteria that you are going to use to select the best model parameters. There are various optimization criteria for that:

- **Mean square error** (MSE)
- **Maximum likelihood** (ML) criterion
- **Maximum a posterior probability** (MAP)

The optimization problem may be hard to solve, but the right choice of model and error function makes a difference.

Evaluating your model

In this step, we try to measure the generalization error of our model on the unseen data. Since we only have the specific data without knowing any unseen data beforehand, we can randomly select a test set from the data and never use it in the training process so that it acts like valid unseen data. There are different ways you can to evaluate the performance of the selected model:

- Simple holdout method, which is dividing the data into training and testing sets
- Other complex methods, based on cross-validation and random subsampling

Our objective in this step is to compare the predictive performance for different models trained on the same data and choose the one with a better (smaller) testing error, which will give us a better generalization error over the unseen data. You can also be more certain about the generalization error by using a statistical method to test the significance of your results.

Getting to learn

Building a machine learning system comes with some challenges and issues; we will try to address them in this section. Many of these issues are domain specific and others aren't.

Challenges of learning

The following is an overview of the challenges and issues that you will typically face when trying to build a learning system.

Feature extraction – feature engineering

Feature extraction is one of the crucial steps toward building a learning system. If you did a good job in this challenge by selecting the proper/right number of features, then the rest of the learning process will be easy. Also, feature extraction is domain dependent and it requires prior knowledge to have a sense of what features could be important for a particular task. For example, the features for our fish recognition system will be different from the ones for spam detection or identifying fingerprints.

The feature extraction step starts from the raw data that you have. Then build derived variables/values (features) that are informative about the learning task and facilitate the next steps of learning and evaluation (generalization).

Some tasks will have a vast number of features and fewer training samples (observations) to facilitate the subsequent learning and generalization processes. In such cases, data scientists use dimensionality reduction techniques to reduce the vast number of features to a smaller set.

Noise

In the fish recognition task, you can see that the length, weight, fish color, as well as the boat color may vary, and there could be shadows, images with low resolution, and other objects in the image. All these issues affect the significance of the proposed explanatory features that should be informative about our fish classification task.

Work-arounds will be helpful in this case. For example, someone might think of detecting the boat ID and mask out certain parts of the boat that most likely won't contain any fish to be detected by our system. This work-around will limit our search space.

Overfitting

As we have seen in our fish recognition task, we have tried to enhance our model's performance by increasing the model complexity and perfectly classifying every single instance of the training samples. As we will see later, such models do not work over unseen data (such as the data that we will use for testing the performance of our model). Having trained models that work perfectly over the training samples but fail to perform well over the testing samples is called **overfitting**.

If you sift through the latter part of the chapter, we build a learning system with an objective to use the training samples as a knowledge base for our model in order to learn from it and generalize over the unseen data. Performance error of the trained model is of no interest to us over the training data; rather, we are interested in the performance (generalization) error of the trained model over the testing samples that haven't been involved in the training phase.

Selection of a machine learning algorithm

Sometimes you are unsatisfied with the execution of the model that you have utilized for a particular errand and you need an alternate class of models. Each learning strategy has its own presumptions about the information it will utilize as a learning base. As an information researcher, you have to discover which suspicions will fit your information best; by this you will have the capacity to acknowledge to attempt a class of models and reject another.

Prior knowledge

As discussed in the concepts of model selection and feature extraction, the two issues can be dealt with, if you have prior knowledge about:

- The appropriate feature
- Model selection parts

Having prior knowledge of the explanatory features in the fish recognition system enabled us to differentiate amid different types of fish. We can go promote by endeavoring to envision our information and get some feeling of the information types of the distinctive fish classifications. On the basis of this prior knowledge, apt family of models can be chosen.

Missing values

Missing features mainly occur because of a lack of data or choosing the prefer-not-to-tell option. How can we handle such a case in the learning process? For example, imagine we find the width of specific a fish type is missing for some reason. There are many ways to handle these missing features.

Implementing the fish recognition/detection model

To introduce the power of machine learning and deep learning in particular, we are going to implement the fish recognition example. No understanding of the inner details of the code will be required. The point of this section is to give you an overview of a typical machine learning pipeline.

Our knowledge base for this task will be a bunch of images, each one of them is labeled as opah or tuna. For this implementation, we are going to use one of the deep learning architectures that made a breakthrough in the area of imaging and computer vision in general. This architecture is called **Convolution Neural Networks** (**CNNs**). It is a family of deep learning architectures that use the convolution operation of image processing to extract features from the images that can explain the object that we want to classify. For now, you can think of it as a magic box that will take our images, learn from it how to distinguish between our two classes (opah and tuna), and then we will test the learning process of this box by feeding it with unlabeled images and see if it's able to tell which type of fish is in the image.

 Different types of learning will be addressed in a later section, so you will understand later on why our fish recognition task is under the supervised learning category.

In this example, we will be using Keras. For the moment, you can think of Keras as an API that makes building and using deep learning way easier than usual. So let's get started! From the Keras website we have:

> *Keras is a high-level neural networks API, written in Python and capable of running on top of TensorFlow, CNTK, or Theano. It was developed with a focus on enabling fast experimentation. Being able to go from idea to result with the least possible delay is key to doing good research.*

Knowledge base/dataset

As we mentioned earlier, we need a historical base of data that will be used to teach the learning algorithm about the task that it's supposed to do later. But we also need another dataset for testing its ability to perform the task after the learning process. So to sum up, we need two types of datasets during the learning process:

1. The first one is the knowledge base where we have the input data and their corresponding labels such as the fish images and their corresponding labels (opah or tuna). This data will be fed to the learning algorithm to learn from it and try to discover the patterns/trends that will help later on for classifying unlabeled images.
2. The second one is mainly for testing the ability of the model to apply what it learned from the knowledge base to unlabeled images or unseen data, in general, and see if it's working well.

As you can see, we only have the data that we will use as a knowledge base for our learning method. All of the data we have at hand will have the correct output associated with it. So we need to somehow make up this data that does not have any correct output associated with it (the one that we are going to apply the model to).

While performing data science, we'll be doing the following:

- **Training phase**: We present our data from our knowledge base and train our learning method/model by feeding the input data along with its correct output to the model.
- **Validation/test phase**: In this phase, we are going to measure how well the trained model is doing. We also use different model property techniques in order to measure the performance of our trained model by using (R-square score for regression, classification errors for classifiers, recall and precision for IR models, and so on).

The validation/test phase is usually split into two steps:

1. In the first step, we use different learning methods/models and choose the best performing one based on our validation data (validation step)
2. Then we measure and report the accuracy of the selected model based on the test set (test step)

Now let's see how we get this data to which we are going to apply the model and see how well trained it is.

Since we don't have any training samples without the correct output, we can make up one from the original training samples that we will be using. So we can split our data samples into three different sets (as shown in *Figure 1.9*):

- **Train set**: This will be used as a knowledge base for our model. Usually, will be 70% from the original data samples.
- **Validation set**: This will be used to choose the best performing model among a set of models. Usually this will be 10% of the original data samples.
- **Test set**: This will be used to measure and report the accuracy of the selected model. Usually, it will be as big as the validation set.

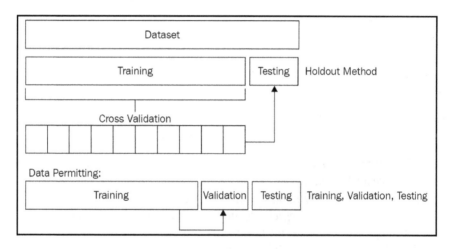

Figure 1.9: Splitting data into train, validation, and test sets

In case you have only one learning method that you are using, you can cancel the validation set and re-split your data to be train and test sets only. Usually, data scientists use 75/25 as percentages, or 70/30.

Data analysis pre-processing

In this section we are going to analyze and preprocess the input images and have it in an acceptable format for our learning algorithm, which is the convolution neural networks here.

So let's start off by importing the required packages for this implementation:

```
import numpy as np
np.random.seed(2018)
```

```
import os
import glob
import cv2
import datetime
import pandas as pd
import time
import warnings
warnings.filterwarnings("ignore")

from sklearn.cross_validation import KFold
from keras.models import Sequential
from keras.layers.core import Dense, Dropout, Flatten
from keras.layers.convolutional import Convolution2D, MaxPooling2D,
ZeroPadding2D
from keras.optimizers import SGD
from keras.callbacks import EarlyStopping
from keras.utils import np_utils
from sklearn.metrics import log_loss
from keras import __version__ as keras_version
```

In order to use the images provided in the dataset, we need to get them to have the same size. OpenCV is a good choice for doing this, from the OpenCV website:

> *OpenCV (Open Source Computer Vision Library) is released under a BSD license and hence it's free for both academic and commercial use. It has C++, C, Python and Java interfaces and supports Windows, Linux, Mac OS, iOS and Android. OpenCV was designed for computational efficiency and with a strong focus on real-time applications. Written in optimized C/C++, the library can take advantage of multi-core processing. Enabled with OpenCL, it can take advantage of the hardware acceleration of the underlying heterogeneous compute platform.*

 You can install OpenCV by using the python package manager by issuing,
`pip install opencv-python`

```
# Parameters
# ----------
# img_path : path
#     path of the image to be resized
def rezize_image(img_path):
    #reading image file
    img = cv2.imread(img_path)
    #Resize the image to to be 32 by 32
    img_resized = cv2.resize(img, (32, 32), cv2.INTER_LINEAR)
return img_resized
```

Now we need to load all the training samples of our dataset and resize each image, according to the previous function. So we are going to implement a function that will load the training samples from the different folders that we have for each fish type:

```
# Loading the training samples and their corresponding labels
def load_training_samples():
    #Variables to hold the training input and output variables
    train_input_variables = []
    train_input_variables_id = []
    train_label = []
    # Scanning all images in each folder of a fish type
    print('Start Reading Train Images')
    folders = ['ALB', 'BET', 'DOL', 'LAG', 'NoF', 'OTHER', 'SHARK', 'YFT']
    for fld in folders:
        folder_index = folders.index(fld)
        print('Load folder {} (Index: {})'.format(fld, folder_index))
        imgs_path = os.path.join('..', 'input', 'train', fld, '*.jpg')
        files = glob.glob(imgs_path)
        for file in files:
            file_base = os.path.basename(file)
            # Resize the image
            resized_img = rezize_image(file)
            # Appending the processed image to the input/output variables of
the classifier
            train_input_variables.append(resized_img)
            train_input_variables_id.append(file_base)
            train_label.append(folder_index)
    return train_input_variables, train_input_variables_id, train_label
```

As we discussed, we have a test set that will act as the unseen data to test the generalization ability of our model. So we need to do the same with testing images; load them and do the resizing processing:

```
def load_testing_samples():
    # Scanning images from the test folder
    imgs_path = os.path.join('..', 'input', 'test_stg1', '*.jpg')
    files = sorted(glob.glob(imgs_path))
    # Variables to hold the testing samples
    testing_samples = []
    testing_samples_id = []
    #Processing the images and appending them to the array that we have
    for file in files:
        file_base = os.path.basename(file)
        # Image resizing
        resized_img = rezize_image(file)
        testing_samples.append(resized_img)
        testing_samples_id.append(file_base)
```

```
    return testing_samples, testing_samples_id
```

Now we need to invoke the previous function into another one that will use the `load_training_samples()` function in order to load and resize the training samples. Also, it will add a few lines of code to convert the training data into NumPy format, reshape that data to fit into our classifier, and finally convert it to float:

```
def load_normalize_training_samples():
    # Calling the load function in order to load and resize the training
samples
    training_samples, training_label, training_samples_id =
load_training_samples()
    # Converting the loaded and resized data into Numpy format
    training_samples = np.array(training_samples, dtype=np.uint8)
    training_label = np.array(training_label, dtype=np.uint8)
    # Reshaping the training samples
    training_samples = training_samples.transpose((0, 3, 1, 2))
    # Converting the training samples and training labels into float format
    training_samples = training_samples.astype('float32')
    training_samples = training_samples / 255
    training_label = np_utils.to_categorical(training_label, 8)
    return training_samples, training_label, training_samples_id
```

We also need to do the same with the test:

```
def load_normalize_testing_samples():
    # Calling the load function in order to load and resize the testing
samples
    testing_samples, testing_samples_id = load_testing_samples()
    # Converting the loaded and resized data into Numpy format
    testing_samples = np.array(testing_samples, dtype=np.uint8)
    # Reshaping the testing samples
    testing_samples = testing_samples.transpose((0, 3, 1, 2))
    # Converting the testing samples into float format
    testing_samples = testing_samples.astype('float32')
    testing_samples = testing_samples / 255
    return testing_samples, testing_samples_id
```

Model building

Now it's time to create the model. As we mentioned, we are going to use a deep learning architecture called CNN as a learning algorithm for this fish recognition task. Again, you are not required to understand any of the previous or the upcoming code in this chapter as we are only demonstrating how complex data science tasks can be solved by using only a few lines of code with the help of Keras and TensorFlow as a deep learning platform.

Also note that CNN and other deep learning architectures will be explained in greater detail in later chapters:

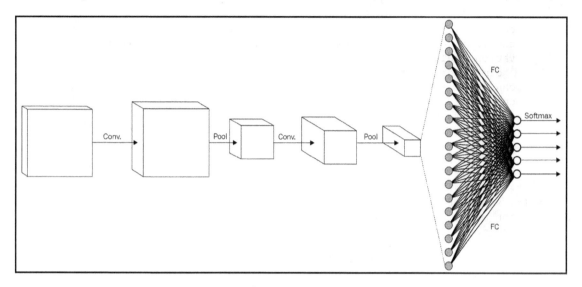

Figure 1.10: CNN architecture

So let's go ahead and create a function that will be responsible for creating the CNN architecture that will be used in our fish recognition task:

```
def create_cnn_model_arch():
    pool_size = 2 # we will use 2x2 pooling throughout
    conv_depth_1 = 32 # we will initially have 32 kernels per conv.
layer...
    conv_depth_2 = 64 # ...switching to 64 after the first pooling layer
    kernel_size = 3 # we will use 3x3 kernels throughout
    drop_prob = 0.5 # dropout in the FC layer with probability 0.5
    hidden_size = 32 # the FC layer will have 512 neurons
    num_classes = 8 # there are 8 fish types
    # Conv [32] -> Conv [32] -> Pool
    cnn_model = Sequential()
    cnn_model.add(ZeroPadding2D((1, 1), input_shape=(3, 32, 32),
dim_ordering='th'))
    cnn_model.add(Convolution2D(conv_depth_1, kernel_size, kernel_size,
activation='relu',
        dim_ordering='th'))
    cnn_model.add(ZeroPadding2D((1, 1), dim_ordering='th'))
    cnn_model.add(Convolution2D(conv_depth_1, kernel_size, kernel_size,
activation='relu',
        dim_ordering='th'))
```

```
    cnn_model.add(MaxPooling2D(pool_size=(pool_size, pool_size),
strides=(2, 2),
      dim_ordering='th'))
    # Conv [64] -> Conv [64] -> Pool
    cnn_model.add(ZeroPadding2D((1, 1), dim_ordering='th'))
    cnn_model.add(Convolution2D(conv_depth_2, kernel_size, kernel_size,
activation='relu',
      dim_ordering='th'))
    cnn_model.add(ZeroPadding2D((1, 1), dim_ordering='th'))
    cnn_model.add(Convolution2D(conv_depth_2, kernel_size, kernel_size,
activation='relu',
      dim_ordering='th'))
    cnn_model.add(MaxPooling2D(pool_size=(pool_size, pool_size),
strides=(2, 2),
      dim_ordering='th'))
    # Now flatten to 1D, apply FC then ReLU (with dropout) and finally
softmax(output layer)
    cnn_model.add(Flatten())
    cnn_model.add(Dense(hidden_size, activation='relu'))
    cnn_model.add(Dropout(drop_prob))
    cnn_model.add(Dense(hidden_size, activation='relu'))
    cnn_model.add(Dropout(drop_prob))
    cnn_model.add(Dense(num_classes, activation='softmax'))
    # initiating the stochastic gradient descent optimiser
    stochastic_gradient_descent = SGD(lr=1e-2, decay=1e-6, momentum=0.9,
nesterov=True)    cnn_model.compile(optimizer=stochastic_gradient_descent,
    # using the stochastic gradient descent optimiser
                 loss='categorical_crossentropy')  # using the cross-
entropy loss function
    return cnn_model
```

Before starting to train the model, we need to use a model assessment and validation method to help us assess our model and see its generalization ability. For this, we are going to use a method called **k-fold cross-validation**. Again, you are not required to understand this method or how it works as we are going to explain this method later in much detail.

So let's start and and create a function that will help us assess and validate the model:

```
def create_model_with_kfold_cross_validation(nfolds=10):
    batch_size = 16 # in each iteration, we consider 32 training examples
at once
    num_epochs = 30 # we iterate 200 times over the entire training set
    random_state = 51 # control the randomness for reproducibility of the
results on the same platform
    # Loading and normalizing the training samples prior to feeding it to
the created CNN model
    training_samples, training_samples_target, training_samples_id =
```

```
    load_normalize_training_samples()
yfull_train = dict()
# Providing Training/Testing indices to split data in the training
samples
# which is splitting data into 10 consecutive folds with shuffling
kf = KFold(len(train_id), n_folds=nfolds, shuffle=True,
random_state=random_state)
fold_number = 0 # Initial value for fold number
sum_score = 0 # overall score (will be incremented at each iteration)
trained_models = [] # storing the modeling of each iteration over the
folds
# Getting the training/testing samples based on the generated
training/testing indices by
    Kfold
for train_index, test_index in kf:
    cnn_model = create_cnn_model_arch()
    training_samples_X = training_samples[train_index] # Getting the
training input variables
    training_samples_Y = training_samples_target[train_index] # Getting
the training output/label variable
    validation_samples_X = training_samples[test_index] # Getting the
validation input variables
    validation_samples_Y = training_samples_target[test_index] # Getting
the validation output/label variable
    fold_number += 1
    print('Fold number {} from {}'.format(fold_number, nfolds))
    callbacks = [
        EarlyStopping(monitor='val_loss', patience=3, verbose=0),
    ]
    # Fitting the CNN model giving the defined settings
    cnn_model.fit(training_samples_X, training_samples_Y,
batch_size=batch_size,
        nb_epoch=num_epochs,
            shuffle=True, verbose=2,
validation_data=(validation_samples_X,
            validation_samples_Y),
            callbacks=callbacks)
    # measuring the generalization ability of the trained model based on
the validation set
    predictions_of_validation_samples =
        cnn_model.predict(validation_samples_X.astype('float32'),
        batch_size=batch_size, verbose=2)
    current_model_score = log_loss(Y_valid,
predictions_of_validation_samples)
    print('Current model score log_loss: ', current_model_score)
    sum_score += current_model_score*len(test_index)
    # Store valid predictions
    for i in range(len(test_index)):
```

```
            yfull_train[test_index[i]] =
predictions_of_validation_samples[i]
        # Store the trained model
        trained_models.append(cnn_model)
    # incrementing the sum_score value by the current model calculated
score
    overall_score = sum_score/len(training_samples)
    print("Log_loss train independent avg: ", overall_score)
    #Reporting the model loss at this stage
    overall_settings_output_string = 'loss_' + str(overall_score) +
'_folds_' + str(nfolds) +
        '_ep_' + str(num_epochs)
    return overall_settings_output_string, trained_models
```

Now, after building the model and using k-fold cross-validation method in order to assess and validate the model, we need to report the results of the trained model over the test set. In order to do this, we are also going to use k-fold cross-validation but this time over the test to see how good our trained model is.

So let's define the function that will take the trained CNN models as an input and then test them using the test set that we have:

```
def test_generality_crossValidation_over_test_set(
overall_settings_output_string, cnn_models):
    batch_size = 16 # in each iteration, we consider 32 training examples
at once
    fold_number = 0 # fold iterator
    number_of_folds = len(cnn_models) # Creating number of folds based on
the value used in the training step
    yfull_test = [] # variable to hold overall predictions for the test set
    #executing the actual cross validation test process over the test set
    for j in range(number_of_folds):
        model = cnn_models[j]
        fold_number += 1
        print('Fold number {} out of {}'.format(fold_number,
number_of_folds))
        #Loading and normalizing testing samples
        testing_samples, testing_samples_id =
load_normalize_testing_samples()
        #Calling the current model over the current test fold
        test_prediction = model.predict(testing_samples,
batch_size=batch_size, verbose=2)
        yfull_test.append(test_prediction)
    test_result = merge_several_folds_mean(yfull_test, number_of_folds)
    overall_settings_output_string = 'loss_' +
overall_settings_output_string \ + '_folds_' +
        str(number_of_folds)
```

```
    format_results_for_types(test_result, testing_samples_id,
  overall_settings_output_string)
```

Model training and testing

Now we are ready to start the model training phase by calling the main function
`create_model_with_kfold_cross_validation()` for building and training the CNN
model using 10-fold cross-validation; then we can call the testing function to measure the
model's ability to generalize to the test set:

```
if __name__ == '__main__':
    info_string, models = create_model_with_kfold_cross_validation()
    test_generality_crossValidation_over_test_set(info_string, models)
```

Fish recognition – all together

After explaining the main building blocks for our fish recognition example, we are ready to
see all the code pieces connected together and see how we managed to build such a
complex system with just a few lines of code. The full code is placed in the *Appendix* section
of the book.

Different learning types

According to Arthur Samuel (`https://en.wikipedia.org/wiki/Arthur_Samuel`), *data
science gives computers the ability to learn without being explicitly programmed*. So, any piece of
software that will consume training examples in order to make decisions over unseen data
without explicit programming is considered learning. Data science or learning comes in
three different forms.

Figure 1.12 shows the commonly used types of data science/machine learning:

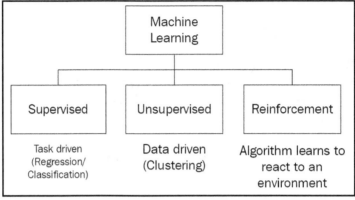

Figure 1.12: Different types of data science/machine learning.

Supervised learning

The majority of data scientists use supervised learning. Supervised learning is where you have some explanatory features, which are called input variables (X), and you have the labels that are associated with the training samples, which are called output variables (Y). The objective of any supervised learning algorithm is to learn the mapping function from the input variables (X) to the output variables (Y):

$$Y = f(X)$$

So the supervised learning algorithm will try to learn approximately the mapping from the input variables (X) to the output variables (Y), such that it can be used later to predict the Y values of an unseen sample.

Figure 1.13 shows a typical workflow for any supervised data science system:

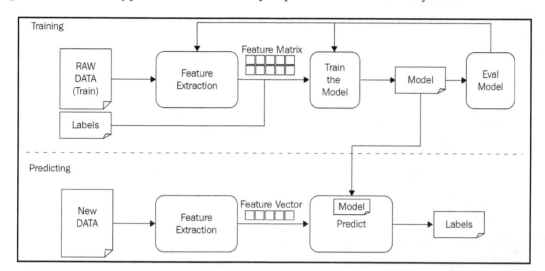

Figure 1.13: A typical supervised learning workflow/pipeline. The top part shows the training process that starts with feeding the raw data into a feature extraction module where we will select meaningful explanatory feature to represent our data. After that, the extracted/selected explanatory feature gets combined with the training set and we feed it to the learning algorithm in order to learn from it. Then we do some model evaluation to tune the parameters and get the learning algorithm to get the best out of the data samples.

This kind of learning is called **supervised learning** because you are getting the label/output of each training sample associated with it. In this case, we can say that the learning process is supervised by a supervisor. The algorithm makes decisions on the training samples and is corrected by the supervisor, based on the correct labels of the data. The learning process will stop when the supervised learning algorithm achieves an acceptable level of accuracy.

Supervised learning tasks come in two different forms; regression and classification:

- **Classification**: A classification task is when the label or the output variable is a category, such as *tuna* or *Opah* or *spam* and *non spam*
- **Regression**: A regression task is when the output variable is a real value, such as *house prices* or *height*

Unsupervised learning

Unsupervised learning is viewed as the second most common kind of learning that is utilized by information researchers. In this type of learning, only the explanatory features or the input variables (X) are given, without any corresponding label or output variable.

The target of unsupervised learning algorithms is to take in the hidden structures and examples in the information. This kind of learning is called **unsupervised** in light of the fact that there aren't marks related with the training samples. So it's a learning procedure without corrections, and the algorithm will attempt to find the basic structure on its own.

Unsupervised learning can be further broken into two forms—clustering and association tasks:

- **Clustering**: A clustering task is where you want to discover similar groups of training samples and group them together, such as grouping documents by topic
- **Association**: An association rule learning task is where you want to discover some rules that describe the relationships in your training samples, such as people who watch movie X also tend to watch movie Y

Figure 1.14 shows a trivial example of unsupervised learning where we have got scattered documents and we are trying to group *similar* ones together:

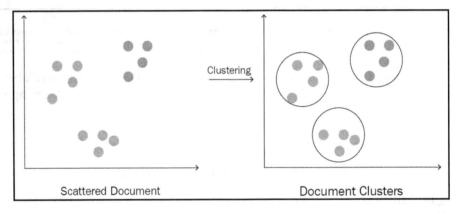

Figure 1.14: Shows how unsupervised use similarity measure such as Euclidean distance to group similar documents to together and draw a decision boundaries for them

Semi-supervised learning

Semi-supervised learning is a type of learning that sits in between supervised and unsupervised learning, where you have got training examples with input variables (X), but only some of them are labeled/tagged with the output variable (Y).

A good example of this type of learning is Flickr (`https://www.flickr.com/`), where you have got lots of images uploaded by users but only some of them are labeled (such as sunset, ocean, and dog) and the rest are unlabeled.

To solve the tasks that fall into this type of learning, you can use one of the following or a combination of them:

- **Supervised learning**: Learn/train the learning algorithm to give predictions about the unlabeled data and then feed the entire training samples back to learn from it and predict the unseen data
- **Unsupervised learning**: Use the unsupervised learning algorithms to learn the underlying structure of the explanatory features or the input variables as if you don't have any tagged training samples

Reinforcement learning

The last type of learning in machine learning is reinforcement learning, in which there's no supervisor but only a reward signal.

So the reinforcement learning algorithm will try to make a decision and then a reward signal will be there to tell whether this decision is right or wrong. Also, this supervision feedback or reward signal may not come instantaneously but get delayed for a few steps. For example, the algorithm will take a decision now, but only after many steps will the reward signal tell whether decision was good or bad.

Data size and industry needs

Data is the information base of our learning calculations; any uplifting and imaginative thoughts will be nothing with the absence of information. So in the event that you have a decent information science application with the right information, at that point you are ready to go.

Having the capacity to investigate and extricate an incentive from your information is obvious these days notwithstanding to the structure of your information, however since enormous information is turning into the watchword of the day then we require information science apparatuses and advancements that can scale with this immense measure of information in an unmistakable learning time. These days everything is producing information and having the capacity to adapt to it is a test. Huge organizations, for example, Google, Facebook, Microsoft, IBM, and so on, manufacture their own adaptable information science arrangements keeping in mind the end goal to deal with the tremendous amount of information being produced once a day by their clients.

TensorFlow, is a machine intelligence/data science platform that was released as an open source library on November 9, 2016 by Google. It is a scalable analytics platform that enables data scientists to build complex systems with a vast amount of data in visible time and it also enables them to use greedy learning methods that require lots of data to get a good performance.

Summary

In this chapter, we went through building a learning system for fish recognition; we also saw how we can build complex applications, such as fish recognition, using a few lines of code with the help of TensorFlow and Keras. This coding example was not meant to be understood from your side, rather to demonstrate the visibility of building complex systems and how data science in general and specifically deep learning became an easy-to-use tool.

We saw the challenges that you might encounter in your daily life as a data scientist while building a learning system.

We also looked at the typical design cycle for building a learning system and explained the overall idea of each component involved in this cycle.

Finally, we went through different learning types, having big data generated daily by big and small companies, and how this vast amount of data raises a red alert to build scalable tools to be able to analyze and extract value from this data.

At this point, the reader may be overwhelmed by all the information mentioned so far, but most of what we explained in this chapter will be addressed in other chapters, including data science challenges and the fish recognition example. The whole purpose of this chapter was to get an overall idea about data science and its development cycle, without any deep understanding of the challenges and the coding example. The coding example was mentioned in this chapter to break the fear of most newcomers in the field of data science and show them how complex systems such as fish recognition can be done in a few lines of code.

Next up, we will start our *by example* journey, by addressing the basic concepts of data science through an example. The next part will mainly focus on preparing you for the later advanced chapters, by going through the famous Titanic example. Lots of concepts will be addressed, including different learning methods for regression and classification, different types of performance errors and which one to care about most, and more about tackling some of the data science challenges and handling different forms of the data samples.

Data Modeling in Action - The Titanic Example

Linear models are the basic learning algorithms in the field of data science. Understanding how a linear model works is crucial in your journey of learning data science because it's the basic building block for most of the sophisticated learning algorithms out there, including neural networks.

In this chapter, we are going to dive into a famous problem in the field of data science, which is the Titanic example. The purpose of this example is to introduce linear models for classification and see a full machine learning system pipeline, starting from data handling and exploration up to model evaluation. We are going to cover the following topics in this chapter:

- Linear models for regression
- Linear models for classification
- Titanic example—model building and training
- Different types of errors

Linear models for regression

Linear regression models are the most basic type of regression models and are widely used in predictive data analysis. The overall idea of regression models is to examine two things:

1. Does a set of explanatory features / input variables do a good job at predicting an output variable? Is the model using features that account for the variability in changes to the dependent variable (output variable)?

2. Which features in particular are significant ones of the dependent variable? And in what way do they impact the dependent variable (indicated by the magnitude and sign of the parameters)? These regression parameters are used to explain the relationship between one output variable (dependent variable) and one or more input features (independent variables).

A regression equation will formulate the impact of the input variables (independent variables) on the output variable (dependent variable). The simplest form of this equation, with one input variable and one output variable, is defined by this formula $y = c + b*x$. Here, y = estimated dependent score, c = constant, b = regression parameter/coefficients, and x = input (independent) variable.

Motivation

Linear regression models are the building blocks of many learning algorithms, but this is not the only reason behind their popularity. The following are the key factors behind their popularity:

- **Widely used**: Linear regression is the oldest regression technique and it's widely used in many applications, such as forecasting and financial analysis.
- **Runs fast**: Linear regression algorithms are very simple and don't include mathematical computations which are too expensive.
- **Easy to use (not a lot of tuning required)**: Linear regression is very easy to use, and mostly it's the first learning method to learn about in the machine learning or data science class as you don't have too many hyperparameters to tune in order to get better performance.
- **Highly interpretable**: Because of its simplicity and ease of inspecting the contribution of each predictor-coefficient pair, linear regression is highly interpretable; you can easily understand the model behavior and interpret the model output for non-technical guys. If a coefficient is zero, the associated predictor variable contributes nothing. If a coefficient is not zero, the contribution due to the specific predictor variable can easily be ascertained.
- **Basis for many other methods**: Linear regression is considered the underlying foundation for many learning methods, such as neural networks and its growing part, deep learning.

Advertising – a financial example

In order to better understand linear regression models, we will go through an example advertisement. We will try to predict the sales of some companies given some factors related to the amount of money spent by these companies on advertising in TV, radio, and newspapers.

Dependencies

To model our advertising data samples using linear regression, we will be using the Stats models library to get nice characteristics for linear models, but later on, we will be using scikit-learn, which has very useful functionality for data science in general.

Importing data with pandas

There are lots of libraries out there in Python that you can use to read, transform, or write data. One of these libraries is pandas (`http://pandas.pydata.org/`). Pandas is an open source library and has great functionality and tools for data analysis as well as very easy-to-use data structures.

You can easily get pandas in many different ways. The best way to get pandas is to install it via `conda` (`http://pandas.pydata.org/pandas-docs/stable/install.html#installing-pandas-with-anaconda`).

> *"conda is an open source package management system and environment management system for installing multiple versions of software packages and their dependencies and switching easily between them. It works on Linux, OS X and Windows, and was created for Python programs but can package and distribute any software." – conda website.*

> *You can easily get conda by installing Anaconda, which is an open data science platform.*

So, let's have a look and see how to use pandas in order to read advertising data samples. First off, we need to import `pandas`:

```
import pandas as pd
```

Next up, we can use the `pandas.read_csv` method in order to load our data into an easy-to-use pandas data structure called **DataFrame**. For more information about `pandas.read_csv` and its parameters, you can refer to the pandas documentation for this method (`https://pandas.pydata.org/pandas-docs/stable/generated/pandas.read_csv.html`):

```
# read advertising data samples into a DataFrame

advertising_data =
pd.read_csv('http://www-bcf.usc.edu/~gareth/ISL/Advertising.csv',
index_col=0)
```

The first argument passed to the `pandas.read_csv` method is a string value representing the file path. The string can be a URL that includes `http`, `ftp`, `s3`, and `file`. The second argument passed is the index of the column that will be used as a label/name for the data rows.

Now, we have the data DataFrame, which contains the advertising data provided in the URL and each row is labeled by the first column. As mentioned earlier, pandas provides easy-to-use data structures that you can use as containers for your data. These data structures have some methods associated with them and you will be using these methods to transform and/or operate on your data.

Now, let's have a look at the first five rows of the advertising data:

```
# DataFrame.head method shows the first n rows of the data where the
# default value of n is 5, DataFrame.head(n=5)
advertising_data.head()
```

Output:

	TV	Radio	Newspaper	Sales
1	230.1	37.8	69.2	22.1
2	44.5	39.3	45.1	10.4
3	17.2	45.9	69.3	9.3
4	151.5	41.3	58.5	18.5
5	180.8	10.8	58.4	12.9

Understanding the advertising data

This problem falls into the supervised learning type, in which we have explanatory features (input variables) and the response (output variable).

What are the features/input variables?

- **TV**: Advertising dollars spent on TV for a single product in a given market (in thousands of dollars)
- **Radio**: Advertising dollars spent on radio
- **Newspaper**: Advertising dollars spent on newspapers

What is the response/outcome/output variable?

- **Sales**: The sales of a single product in a given market (in thousands of widgets)

We can also use the `DataFrame` method shape to know the number of samples/observations in our data:

```
# print the shape of the DataFrame
advertising_data.shape
Output:
(200, 4)
```

So, there are 200 observations in the advertising data.

Data analysis and visualization

In order to understand the underlying form of the data, the relationship between the features and response, and more insights, we can use different types of visualization. To understand the relationship between the advertising data features and response, we are going to use a scatterplot.

In order to make different types of visualizations of your data, you can use Matplotlib (https://matplotlib.org/), which is a Python 2D library for making visualizations. To get Matplotlib, you can follow their installation instructions at: https://matplotlib.org/users/installing.html.

Let's import the visualization library Matplotlib:

```
import matplotlib.pyplot as plt

# The next line will allow us to make inline plots that could appear
```

```
directly in the notebook
# without poping up in a different window
%matplotlib inline
```

Now, let's use a scatterplot to visualize the relationship between the advertising data features and response variable:

```
fig, axs = plt.subplots(1, 3, sharey=True)

# Adding the scatterplots to the grid
advertising_data.plot(kind='scatter', x='TV', y='sales', ax=axs[0],
figsize=(16, 8))
advertising_data.plot(kind='scatter', x='radio', y='sales', ax=axs[1])
advertising_data.plot(kind='scatter', x='newspaper', y='sales', ax=axs[2])
```

Output:

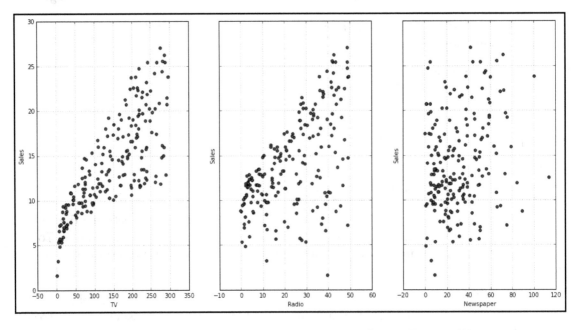

Figure 1: Scatter plot for understanding the relationship between the advertising data features and the response variable

Now, we need to see how the ads will help increase the sales. So, we need to ask ourselves a couple of questions about that. Worthwhile questions to ask will be something like the relationship between the ads and sales, which kind of ads contribute more to the sales, and the approximate effect of each type of ad on the sales. We will try to answer such questions using a simple linear model.

Simple regression model

The linear regression model is a learning algorithm that is concerned with predicting a **quantitative** (also known as **numerical**) **response** using a combination of **explanatory features** (or **inputs** or **predictors**).

A simple linear regression model with only one feature takes the following form:

$$y = beta_0 + beta_1 x$$

Here:

- y is the predicted numerical value (response) → **sales**
- x is the the feature
- $beta_0$ is called the **intercept**
- $beta_1$ is the coefficient of the feature x → **TV ad**

Both $beta_0$ and $beta_1$ are considered as model **coefficients**. In order to create a model that can predict the value of sales in the advertising example, we need to learn these coefficients because $beta_1$ will be the learned effect of the feature x on the response y. For example, if $beta_1 = 0.04$, it means that an additional \$100 spent on TV ads is **associated with** an increase in sales by four widgets. So, we need to go ahead and see how can we learn these coefficients.

Learning model coefficients

In order to estimate the coefficients of our model, we need to fit the data with a regression line that gives a similar answer to the actual sales. To get a regression line that best fits the data, we will use a criterion called **least squares**. So, we need to find a line that minimizes the difference between the predicted value and the observed (actual) one. In other words, we need to find a regression line that minimizes the **sum of squared residuals (SSresiduals)**. *Figure 2* illustrates this:

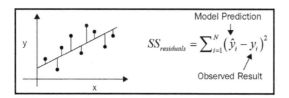

Figure 2: Fitting the data points (sample of TV ads) with a regression line that minimizes the sum of the squared residuals (difference between the predicted and observed value)

The following are the elements that exist in *Figure 2*:

- **Black dots** represent the actual or observed values of x (TV ad) and y (sales)
- **The blue line** represents the least squares line (regression line)
- **The red line** represents the residuals, which are the differences between the predicted and the observed (actual) values

So, this is how our coefficients relate to the least squares line (regression line):

- $beta_0$ is the **intercept,** which is the value of y when $x = 0$
- $beta_1$ is the **slope,** which represents the change in y divided by the change in x

Figure 3 presents a graphical explanation of this:

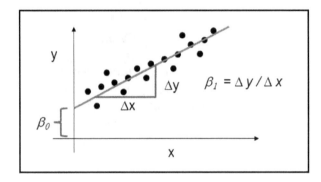

Figure 3: The relation between the least squares line and the model coefficients

Now, let's go ahead and start to learn these coefficients using **Statsmodels**:

```
# To use the formula notation below, we need to import the module like the
following
import statsmodels.formula.api as smf
# create a fitted model in one line of code(which will represent the least
squares line)
lm = smf.ols(formula='sales ~ TV', data=advertising_data).fit()
# show the trained model coefficients
lm.params
```

Output:

```
Intercept       7.032594
TV              0.047537
dtype: float64
```

As we mentioned, one of the advantages of linear regression models is that they are easy to interpret, so let's go ahead and interpret the model.

Interpreting model coefficients

Let's see how to interpret the coefficients of the model, such as the TV ad coefficient (*beta₁*):

- A unit increase in the input/feature (TV ad) spending is **associated** with a `0.047537` unit increase in Sales (response). In other words, an additional $100 spent on TV ads is **associated with** an increase in sales of 4.7537 widgets.

The goal of building a learned model from the TV ad data is to predict the sales for unseen data. So, let's see how we can use the learned model in order to predict the value of sales (which we don't know) based on a given value of a TV ad.

Using the model for prediction

Let's say we have unseen data of TV ad spending and that we want to know their corresponding impact on the sales of the company. So, we need to use the learned model to do that for us. Let's suppose that we want to know how much sales will increase from $50000 of TV advertising.

Let's use our learned model coefficients to make such a calculation:

$$y = 7.032594 + 0.047537 \ x \ 50$$

```
# manually calculating the increase in the sales based on $50k
7.032594 + 0.047537*50000
```

Output:

```
9,409.444
```

We can also use Statsmodels to make the prediction for us. First, we need to provide the TV ad value in a pandas DataFrame since the Statsmodels interface expects it:

```
# creating a Pandas DataFrame to match Statsmodels interface expectations
new_TVAdSpending = pd.DataFrame({'TV': [50000]})
new_TVAdSpending.head()
```

Output:

	TV
0	50000

Now, we can go ahead and use the predict function to predict the sales value:

```
# use the model to make predictions on a new value
preds = lm.predict(new_TVAdSpending)
```

Output:

```
array([ 9.40942557])
```

Let's see how the learned least squares line looks. In order to draw the line, we need two points, with each point represented by this pair: (x, predict_value_of_x).

So, let's take the minimum and maximum values for the TV ad feature:

```
# create a DataFrame with the minimum and maximum values of TV
X_min_max = pd.DataFrame({'TV': [advertising_data.TV.min(),
advertising_data.TV.max()]})
X_min_max.head()
```

Output:

	TV
0	0.7
1	296.4

Let's get the corresponding predictions for these two values:

```
# predictions for X min and max values
predictions = lm.predict(X_min_max)
predictions
```

Output:

```
array([  7.0658692,   21.12245377])
```

Now, let's plot the actual data and then fit it with the least squares line:

```
# plotting the acutal observed data
advertising_data.plot(kind='scatter', x='TV', y='sales')
#plotting the least squares line
plt.plot(new_TVAdSpending, preds, c='red', linewidth=2)
```

Output:

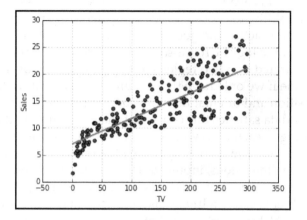

Figure 4: Plot of the actual data and the least squares line

Extensions of this example and further explanations will be explained in the next chapter.

Linear models for classification

In this section, we are going to go through logistic regression, which is one of the widely used algorithms for classification.

What's logistic regression? The simple definition of logistic regression is that it's a type of classification algorithm involving a linear discriminant.

We are going to clarify this definition in two points:

1. Unlike linear regression, logistic regression doesn't try to estimate/predict the value of the numeric variable given a set of features or input variables. Instead, the output of the logistic regression algorithm is the probability that the given sample/observation belongs to a specific class. In simpler words, let's assume that we have a binary classification problem. In this type of problem, we have only two classes in the output variable, for example, diseased or not diseased. So, the probability that a certain sample belongs to the diseased class is P_0 and the probability that a certain sample belongs to the not diseased class is $P_1 = 1 - P_0$. Thus, the output of the logistic regression algorithm is always between 0 and 1.

2. As you probably know, there are a lot of learning algorithms for regression or classification, and each learning algorithm has its own assumption about the data samples. The ability to choose the learning algorithm that fits your data will come gradually with practice and good understanding of the subject. Thus, the central assumption of the logistic regression algorithm is that our input/feature space could be separated into two regions (one for each class) by a linear surface, which could be a line if we only have two features or a plane if we have three, and so on. The position and orientation of this boundary will be determined by your data. If your data satisfies this constraint that is separating them into regions corresponding to each class with a linear surface, then your data is said to be linearly separable. The following figure illustrates this assumption. In *Figure 5*, we have three dimensions, inputs, or features and two possible classes: diseased (red) and not diseased (blue). The dividing place that separates the two regions from each other is called a **linear discriminant**, and that's because it's linear and it helps the model to discriminate between samples belonging to different classes:

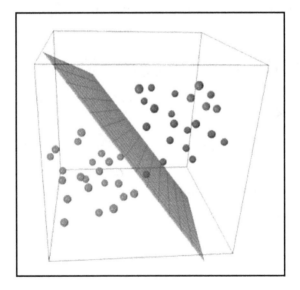

Figure 5: Linear decision surface separating two classes

If your data samples aren't linearly separable, you can make them so by transforming your data into higher dimensional space, by adding more features.

Classification and logistic regression

In the previous section, we learned how to predict continuous quantities (for example, the impact of TV advertising on company sales) as linear functions of input values (for example, TV, Radio, and newspaper advertisements). But for other tasks, the output will not be continuous quantities. For example, predicting whether someone is diseased or not is a classification problem and we need a different learning algorithm to perform this. In this section, we are going to dig deeper into the mathematical analysis of logistic regression, which is a learning algorithm for classification tasks.

In linear regression, we tried to predict the value of the output variable $y^{(i)}$ for the ith sample $x^{(i)}$ in that dataset using a linear model function $y = h_\theta(x) = \theta^T x$. This is not really a great solution for classification tasks such as predicting binary labels ($y^{(i)} \in \{0,1\}$).

Logistic regression is one of the many learning algorithms that we can use for classification tasks, whereby we use a different hypothesis class while trying to predict the probability that a specific sample belongs to the one class and the probability that it belongs to the zero class. So, in logistic regression, we will try to learn the following functions:

$$P(y = 1|x) = h_\theta(x) = \frac{1}{1 + \exp(-\theta^T x)} \equiv \sigma(\theta^T x),$$
$$P(y = 0|x) = 1 - P(y = 1|x) = 1 - h_\theta(x).$$

The function $\sigma(z) \equiv \frac{1}{1 + \exp(-z)}$ is often called a **sigmoid** or **logistic** function, which squashes the value of $\theta^T x$ into a fixed range [0,1], as shown in the following graph. Because the value will be squashed between [0,1], we can then interpret $h_\theta(x)$ as a probability.

Our goal is to search for a value of the parameters θ so that the probability $P(y = 1|x) = h_\theta(x)$ is large when the input sample x belongs to the one class and small when x belongs to the zero class:

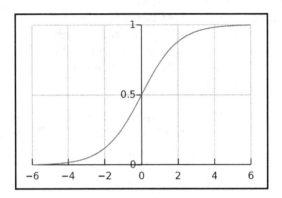

Figure 6: Shape of the sigmoid function

So, suppose we have a set of training samples with their corresponding binary labels $\{(x^{(i)}, y^{(i)}): i = 1,...,m\}$. We will need to minimize the following cost function, which measures how good a given h_θ does:

$$J(\theta) = -\sum_i \left(y^{(i)} \log(h_\theta(x^{(i)})) + (1 - y^{(i)}) \log(1 - h_\theta(x^{(i)})) \right).$$

Note that we have only one of the two terms of the equation's summation as non-zero for each training sample (depending on whether the value of the label $y^{(i)}$ is 0 or). When $y^{(i)} = 1$, minimizing the model cost function means we need to make $h_\theta(x^{(i)})$ large, and when $y^{(i)} = 0$

$$y^{(i)} = 0$$, we want to make $1-h_\theta$ large.

Now, we have a cost function that calculates how well a given hypothesis h_θ fits our training samples. We can learn to classify our training samples by using an optimization technique to minimize $J(\theta)$ and find the best choice of parameters θ. Once we have done this, we can use these parameters to classify a new test sample as 1 or 0, checking which of these two class labels is most probable. If $P(y = 1|x) < P(y = 0|x)$ then we output 0, otherwise we output 1, which is the same as defining a threshold of 0.5 between our classes and checking whether $h_\theta(x) > 0.5$.

To minimize the cost function $J(\theta)$, we can use an optimization technique that finds the best value of θ that minimizes the cost function. So, we can use a calculus tool called **gradient**, which tries to find the greatest rate of increase of the cost function. Then, we can take the opposite direction to find the minimum value of this function; for example, the gradient of $J(\theta)$ is denoted by $\nabla_\theta J(\theta)$, which means taking the gradient for the cost function with respect to the model parameters. Thus, we need to provide a function that computes $J(\theta)$ and $\nabla_\theta J(\theta)$ for any requested choice of θ. If we derived the gradient or derivative of the cost function above $J(\theta)$ with respect to θ_j, we will get the following results:

$$\frac{\partial J(\theta)}{\partial \theta_j} = \sum_i x_j^{(i)} \left(h_\theta(x^{(i)}) - y^{(i)}\right).$$

Which can be written in a vector form as:

$$\nabla_\theta J(\theta) = \sum_i x^{(i)} \left(h_\theta(x^{(i)}) - y^{(i)}\right)$$

Now, we have a mathematical understanding of the logistic regression, so let's go ahead and use this new learning method for solving a classification task.

Titanic example – model building and training

The sinking of the ship, Titanic, is one of the most infamous events in history. This incident led to the deaths of 1,502 passengers and crew out of 2,224. In this problem, we will use data science to predict whether the passenger will survive this tragedy or not and then test the performance of our model based on the actual statistics of the tragedy.

To follow up with the Titanic example, you need to do the following:

1. Download this repository in a ZIP file by clicking on `https://github.com/ahmed-menshawy/ML_Titanic/archive/master.zip` or execute from the terminal:

2. Git clone: `https://github.com/ahmed-menshawy/ML_Titanic.git`

3. Install `[virtualenv]`: (`http://virtualenv.readthedocs.org/en/latest/installation.html`)

4. Navigate to the directory where you unzipped or cloned the repo and create a virtual environment with `virtualenv ml_titanic`

5. Activate the environment with `source ml_titanic/bin/activate`

6. Install the required dependencies with `pip install -r requirements.txt`

7. Execute the `ipython notebook` from the command line or terminal

8. Follow the example code in the chapter

9. When you're done, deactivate the virtual environment with `deactivate`

Data handling and visualization

In this section, we are going to do some data preprocessing and analysis. Data exploration and analysis is considered one of the most important steps while applying machine learning and might also be considered as the most important one, because at this step, you get to know the friend, Data, which is going to stick with you during the training process. Also, knowing your data will enable you to narrow down the set of candidate algorithms that you might use to check which one is the best for your data.

Let's start off by importing the necessary packages for our implementation:

```
import matplotlib.pyplot as plt
%matplotlib inline

from statsmodels.nonparametric.kde import KDEUnivariate
from statsmodels.nonparametric import smoothers_lowess
from pandas import Series, DataFrame
from patsy import dmatrices
from sklearn import datasets, svm

import numpy as np
import pandas as pd
import statsmodels.api as sm

from scipy import stats
stats.chisqprob = lambda chisq, df: stats.chi2.sf(chisq, df)
```

Let's read the Titanic passengers and crew data using Pandas:

```
titanic_data = pd.read_csv("data/titanic_train.csv")
```

Next up, let's check the dimensions of our dataset and see how many examples we have and how many explanatory features are describing our dataset:

```
titanic_data.shape

Output:
(891, 12)
```

So, we have a total of 891 observations, data samples, or passenger/crew records, and 12 explanatory features for describing this record:

```
list(titanic_data)

Output:
['PassengerId',
 'Survived',
 'Pclass',
 'Name',
 'Sex',
 'Age',
 'SibSp',
 'Parch',
 'Ticket',
 'Fare',
 'Cabin',
 'Embarked']
```

Let's see the data of some samples/observations:

```
titanic_data[500:510]
```

Output:

	PassengerId	Survived	Pclass	Name	Sex	Age	SibSp	Parch	Ticket	Fare	Cabin	Embarked
500	501	0	3	Calic, Mr. Petar	male	17.0	0	0	315086	8.6625	NaN	S
501	502	0	3	Canavan, Miss. Mary	female	21.0	0	0	364846	7.7500	NaN	Q
502	503	0	3	O'Sullivan, Miss. Bridget Mary	female	NaN	0	0	330909	7.6292	NaN	Q
503	504	0	3	Laitinen, Miss. Kristina Sofia	female	37.0	0	0	4135	9.5875	NaN	S
504	505	1	1	Maioni, Miss. Roberta	female	16.0	0	0	110152	86.5000	B79	S
505	506	0	1	Penasco y Castellana, Mr. Victor de Satode	male	18.0	1	0	PC 17758	108.9000	C65	C
506	507	1	2	Quick, Mrs. Frederick Charles (Jane Richards)	female	33.0	0	2	26360	26.0000	NaN	S
507	508	1	1	Bradley, Mr. George ("George Arthur Brayton")	male	NaN	0	0	111427	26.5500	NaN	S
508	509	0	3	Olsen, Mr. Henry Margido	male	28.0	0	0	C 4001	22.5250	NaN	S
509	510	1	3	Lang, Mr. Fang	male	26.0	0	0	1601	56.4958	NaN	S

Figure 7: Samples from the titanic dataset

Now, we have a Pandas DataFrame that holds the information of 891 passengers that we need to analyze. The columns of the DataFrame represent the explanatory features about each passenger/crew, like name, sex, or age.

Some of these explanatory features are complete without any missing values, such as the survived feature, which has 891 entries. Other explanatory features contain missing values, such as the age feature, which has only 714 entries. Any missing value in the DataFrame is represented as NaN.

If you explore all of the dataset features, you will find that the ticket and cabin features have many missing values (NaNs), and so they won't add much value to our analysis. To handle this, we will drop them from the DataFrame.

Use the following line of code to drop the `ticket` and `cabin` features entirely from the DataFrame:

```
titanic_data = titanic_data.drop(['Ticket','Cabin'], axis=1)
```

There are a lot of reasons to have such missing values in our dataset. But in order to preserve the integrity of the dataset, we need to handle such missing values. In this specific problem, we will choose to drop them.

Use the following line of code in order to remove all NaN values from all the remaining features:

```
titanic_data = titanic_data.dropna()
```

Now, we have a sort of compete dataset that we can use to do our analysis. If you decided to just delete all the NaNs without deleting the **ticket** and **cabin** features first, you will find that most of the dataset is removed, because the `.dropna()` method removes an observation from the DataFrame, even if it has only one NaN in one of the features.

Let's do some data visualization to see the distribution of some features and understand the relationship between the explanatory features:

```
# declaring graph parameters
fig = plt.figure(figsize=(18,6))
alpha=alpha_scatterplot = 0.3
alpha_bar_chart = 0.55
# Defining a grid of subplots to contain all the figures
ax1 = plt.subplot2grid((2,3),(0,0))
# Add the first bar plot which represents the count of people who survived
vs not survived.
titanic_data.Survived.value_counts().plot(kind='bar',
alpha=alpha_bar_chart)
# Adding margins to the plot
ax1.set_xlim(-1, 2)
# Adding bar plot title
plt.title("Distribution of Survival, (1 = Survived)")
plt.subplot2grid((2,3),(0,1))
plt.scatter(titanic_data.Survived, titanic_data.Age,
alpha=alpha_scatterplot)
# Setting the value of the y label (age)
plt.ylabel("Age")
# formatting the grid
plt.grid(b=True, which='major', axis='y')
plt.title("Survival by Age, (1 = Survived)")
ax3 = plt.subplot2grid((2,3),(0,2))
titanic_data.Pclass.value_counts().plot(kind="barh", alpha=alpha_bar_chart)
ax3.set_ylim(-1, len(titanic_data.Pclass.value_counts()))
plt.title("Class Distribution")
plt.subplot2grid((2,3),(1,0), colspan=2)
# plotting kernel density estimate of the subse of the 1st class passenger's
age
titanic_data.Age[titanic_data.Pclass == 1].plot(kind='kde')
titanic_data.Age[titanic_data.Pclass == 2].plot(kind='kde')
titanic_data.Age[titanic_data.Pclass == 3].plot(kind='kde')
# Adding x label (age) to the plot
plt.xlabel("Age")
```

```
plt.title("Age Distribution within classes")
# Add legend to the plot.
plt.legend(('1st Class', '2nd Class','3rd Class'),loc='best')
ax5 = plt.subplot2grid((2,3),(1,2))
titanic_data.Embarked.value_counts().plot(kind='bar',
alpha=alpha_bar_chart)
ax5.set_xlim(-1, len(titanic_data.Embarked.value_counts()))
plt.title("Passengers per boarding location")
```

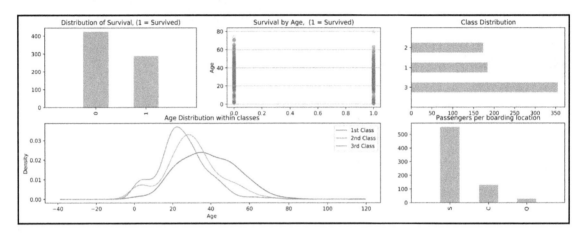

Figure 8: Basic visualizations for the Titanic data samples

As we mentioned, the purpose of this analysis is to predict if a specific passenger will survive the tragedy based on the available feature, such as traveling class (called `pclass` in the data), **Sex**, **Age**, and **Fare Price**. So, let's see if we can get a better visual understanding of the passengers who survived and died.

First, let's draw a bar plot to see the number of observations in each class (survived/died):

```
plt.figure(figsize=(6,4))
fig, ax = plt.subplots()
titanic_data.Survived.value_counts().plot(kind='barh', color="blue",
alpha=.65)
ax.set_ylim(-1, len(titanic_data.Survived.value_counts()))
plt.title("Breakdown of survivals(0 = Died, 1 = Survived)")
```

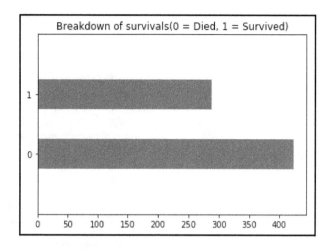

Figure 9: Survival breakdown

Let's get some more understanding of the data by breaking down the previous graph by gender:

```
fig = plt.figure(figsize=(18,6))
#Plotting gender based analysis for the survivals.
male = titanic_data.Survived[titanic_data.Sex ==
'male'].value_counts().sort_index()
female = titanic_data.Survived[titanic_data.Sex ==
'female'].value_counts().sort_index()
ax1 = fig.add_subplot(121)
male.plot(kind='barh',label='Male', alpha=0.55)
female.plot(kind='barh', color='#FA2379',label='Female', alpha=0.55)
plt.title("Gender analysis of survivals (raw value counts) ");
plt.legend(loc='best')
ax1.set_ylim(-1, 2)
ax2 = fig.add_subplot(122)
(male/float(male.sum())).plot(kind='barh',label='Male', alpha=0.55)
(female/float(female.sum())).plot(kind='barh',
color='#FA2379',label='Female', alpha=0.55)
plt.title("Gender analysis of survivals"); plt.legend(loc='best')
ax2.set_ylim(-1, 2)
```

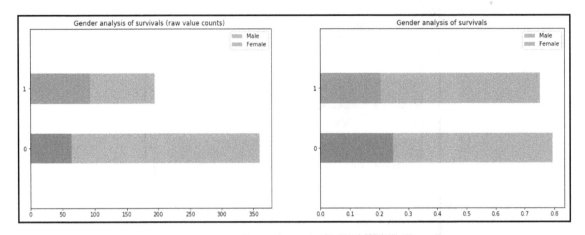

Figure 10: Further breakdown for the Titanic data by the gender feature

Now, we have more information about the two possible classes (survived and died). The exploration and visualization step is necessary because it gives you more insight into the structure of the data and helps you to choose the suitable learning algorithm for your problem. As you can see, we started with very basic plots and then increased the complexity of the plot to discover more about the data that we were working with.

Data analysis – supervised machine learning

The purpose of this analysis is to predict the survivors. So, the outcome will be survived or not, which is a binary classification problem; in it, you have only two possible classes.

There are lots of learning algorithms that we can use for binary classification problems. Logistic regression is one of them. As explained by Wikipedia:

> *In statistics, logistic regression or logit regression is a type of regression analysis used for predicting the outcome of a categorical dependent variable (a dependent variable that can take on a limited number of values, whose magnitudes are not meaningful but whose ordering of magnitudes may or may not be meaningful) based on one or more predictor variables. That is, it is used in estimating empirical values of the parameters in a qualitative response model. The probabilities describing the possible outcomes of a single trial are modeled, as a function of the explanatory (predictor) variables, using a logistic function. Frequently (and subsequently in this article) "logistic regression" is used to refer specifically to the problem in which the dependent variable is binary—that is, the number of available categories is two—and problems with more than two categories are referred to as multinomial logistic regression or, if the multiple categories are ordered, as ordered logistic regression. Logistic regression measures the relationship between a categorical dependent variable and one or more independent variables, which are usually (but not necessarily) continuous, by using probability scores as the predicted values of the dependent variable.[1] As such it treats the same set of problems as does probit regression using similar techniques.*

In order to use logistic regression, we need to create a formula that tells our model the type of features/inputs we're giving it:

```
# model formula
# here the ~ sign is an = sign, and the features of our dataset
# are written as a formula to predict survived. The C() lets our
# regression know that those variables are categorical.
# Ref: http://patsy.readthedocs.org/en/latest/formulas.html
formula = 'Survived ~ C(Pclass) + C(Sex) + Age + SibSp + C(Embarked)'
# create a results dictionary to hold our regression results for easy
analysis later
results = {}
# create a regression friendly dataframe using patsy's dmatrices function
y,x = dmatrices(formula, data=titanic_data, return_type='dataframe')
# instantiate our model
model = sm.Logit(y,x)
# fit our model to the training data
res = model.fit()
# save the result for outputing predictions later
results['Logit'] = [res, formula]
res.summary()
Output:
Optimization terminated successfully.
```

```
Current function value: 0.444388
Iterations 6
```

Logit Regression Results			
Dep. Variable:	Survived	No. Observations:	712
Model:	Logit	Df Residuals:	704
Method:	MLE	Df Model:	7
Date:	Sun, 20 Dec 2015	Pseudo R-squ.:	0.3414
Time:	11:27:33	Log-Likelihood:	-316.40
converged:	True	LL-Null:	-480.45
		LLR p-value:	5.992e-67

	coef	std err	z	P>|z|	[95.0% Conf. Int.]
Intercept	4.5423	0.474	9.583	0.000	3.613 5.471
C(Pclass)[T.2]	-1.2673	0.299	-4.245	0.000	-1.852 -0.682
C(Pclass)[T.3]	-2.4966	0.296	-8.422	0.000	-3.078 -1.916
C(Sex)[T.male]	-2.6239	0.218	-12.060	0.000	-3.050 -2.197
C(Embarked)[T.Q]	-0.8351	0.597	-1.398	0.162	-2.006 0.335
C(Embarked)[T.S]	-0.4254	0.271	-1.572	0.116	-0.956 0.105
Age	-0.0436	0.008	-5.264	0.000	-0.060 -0.027
SibSp	-0.3697	0.123	-3.004	0.003	-0.611 -0.129

Figure 11: Logistic regression results

Now, let's plot the prediction of our model versus actual ones and also the residuals, which is the difference between the actual and predicted values of the target variable:

```
# Plot Predictions Vs Actual
plt.figure(figsize=(18,4));
plt.subplot(121, axisbg="#DBDBDB")
# generate predictions from our fitted model
ypred = res.predict(x)
plt.plot(x.index, ypred, 'bo', x.index, y, 'mo', alpha=.25);
plt.grid(color='white', linestyle='dashed')
plt.title('Logit predictions, Blue: \nFitted/predicted values: Red');
# Residuals
ax2 = plt.subplot(122, axisbg="#DBDBDB")
plt.plot(res.resid_dev, 'r-')
plt.grid(color='white', linestyle='dashed')
ax2.set_xlim(-1, len(res.resid_dev))
plt.title('Logit Residuals');
```

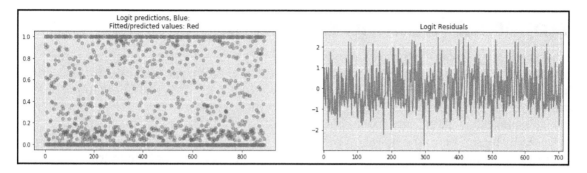

Figure 12: Understanding the logit regression model

Now, we have built our logistic regression model, and prior to that, we have done some analysis and exploration of the dataset. The preceding example shows you the general pipelines for building a machine learning solution.

Most of the time, practitioners fall into some technical pitfalls because they lack experience of understanding the concepts of machine learning. For example, someone might get an accuracy of 99% over the test set, and then without doing any investigation of the distribution of classes in the data (such as how many samples are negative and how many samples are positive), they deploy the model.

To highlight some of these concepts and differentiate between different kinds of errors that you need to be aware of and which ones you should really care about, we'll move on to the next section.

Different types of errors

In machine learning, there are two types of errors, and as a newcomer to data science, you need to understand the crucial difference between both of them. If you end up minimizing the wrong type of error, the whole learning system will be useless and you won't be able to use it in practice over unseen data. To minimize this kind of misunderstanding between practitioners about these two types of errors, we are going to explain them in the following two sections.

Apparent (training set) error

This the first type of error that you don't have to care about minimizing. Getting a small value for this type of error doesn't mean that your model will work well over the unseen data (generalize). To better understand this type of error, we'll give a trivial example of a class scenario. The purpose of solving problems in the classroom is not to be able to solve the same problem again in the exam, but to be able to solve other problems that won't necessarily be similar to the ones you practiced in the classroom. The exam problems could be from the same family of the classroom problems, but not necessarily identical.

Apparent error is the ability of the trained model to perform on the training set for which we already know the true outcome/output. If you manage to get 0 error over the training set, then it is a good indicator for you that your model (mostly) won't work well on unseen data (won't generalize). On the other hand, data science is about using a training set as a base knowledge for the learning algorithm to work well on future unseen data.

In *Figure 3*, the red curve represents the **apparent** error. Whenever you increase the model's ability to memorize things (such as increasing the model complexity by increasing the number of explanatory features), you will find that this apparent error approaches zero. It can be shown that if you have as many features as observations/samples, then the **apparent** error will be zero:

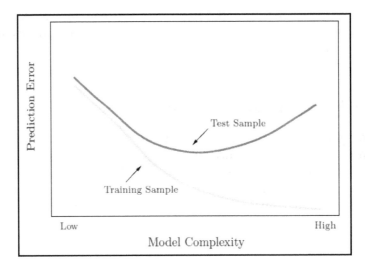

Figure 13: Apparent error (red curve) and generalization/true error (light blue)

Generalization/true error

This is the second and more important type of error in data science. The whole purpose of building learning systems is the ability to get a smaller generalization error on the test set; in other words, to get the model to work well on a set of observation/samples that haven't been used in the training phase. If you still consider the class scenario from the previous section, you can think of generalization error as the ability to solve exam problems that weren't necessarily similar to the problems you solved in the classroom to learn and get familiar with the subject. So, generalization performance is the model's ability to use the skills (parameters) that it learned in the training phase in order to correctly predict the outcome/output of unseen data.

In *Figure 13*, the light blue line represents the generalization error. You can see that as you increase the model complexity, the generalization error will be reduced, until some point when the model will start to lose its increasing power and the generalization error will decrease. This part of the curve where you get the generalization error to lose its increasing generalization power, is called **overfitting**.

The takeaway message from this section is to minimize the generalization error as much as you can.

Summary

A linear model is a very powerful tool that you can use as an initial learning algorithm if your data matches its assumptions. Understanding linear models will help you to understand more sophisticated models that use linear models as building blocks.

Next up, we will continue using the Titanic example by addressing model complexity and assessment in more detail. Model complexity is a very powerful tool and you need to use it carefully in order to enhance the generalization error. Misunderstanding it will lead to overfitting problems.

3
Feature Engineering and Model Complexity – The Titanic Example Revisited

Model complexity and assessment is a must-do step toward building a successful data science system. There are lots of tools that you can use to assess and choose your model. In this chapter, we are going to address some of the tools that can help you to increase the value of your data by adding more descriptive features and extracting meaningful information from existing ones. We are also going to address other tools related optimal number features and learn why it's a problem to have a large number of features and fewer training samples/observations.

The following are the topics that will be explained in this chapter:

- Feature engineering
- The curse of dimensionality
- Titanic example revisited—all together
- Bias-variance decomposition
- Learning visibility

Feature engineering

Feature engineering is one of the key components that contribute to the model's performance. A simple model with the right features can perform better than a complicated one with poor features. You can think of the feature engineering process as the most important step in determining your predictive model's success or failure. Feature engineering will be much easier if you understand the data.

Feature engineering is used extensively by anyone who uses machine learning to solve only one question, which is: **how do you get the most out of your data samples for predictive modeling**? This is the problem that the process and practice of feature engineering solves, and the success of your data science skills starts by knowing how to represent your data well.

Predictive modeling is a formula or rule that transforms a list of features or input variables $(x_1, x_2,..., x_n)$ into an output/target of interest (y). So, what is feature engineering? It's the process of creating new input variables or features $(z_1, z_2, ..., z_n)$ from existing input variables $(x_1, x_2,..., x_n)$. We don't just create any new features; the newly created features should contribute and be relevant to the model's output. Creating such features that will be relevant to the model's output will be an easy process with knowledge of the domain (such as marketing, medical, and so on). Even if machine learning practitioners interact with some domain experts during this process, the outcome of the feature engineering process will be much better.

An example where domain knowledge can be helpful is modeling the likelihood of rain, given a set of input variables/features (temperature, wind speed, and percentage of cloud cover). For this specific example, we can construct a new binary feature called **overcast**, where its value equals 1 or no whenever the percentage of cloud cover is less than 20%, and equals 0 or yes otherwise. In this example, domain knowledge was essential to specify the threshold or cut-off percentage. The more thoughtful and useful the inputs, the better the reliability and predictivity of your model.

Types of feature engineering

Feature engineering as a technique has three main subcategories. As a deep learning practitioner, you have the freedom to choose between them or combine them in some way.

Feature selection

Sometimes called **feature importance**, this is the process of ranking the input variables according to their contribution to the target/output variable. Also, this process can be considered a ranking process of the input variables according to their value in the predictive ability of the model.

Some learning methods do this kind of feature ranking or importance as part of their internal procedures (such as decision trees). Mostly, these kind of methods uses entropy to filter out the less valuable variables. In some cases, deep learning practitioners use such learning methods to select the most important features and then feed them into a better learning algorithm.

Dimensionality reduction

Dimensionality reduction is sometimes feature extraction, and it is the process of combining the existing input variables into a new set of a much reduced number of input variables. One of the most used methods for this type of feature engineering is **principle component analysis (PCA)**, which utilizes the variance in data to come up with a reduced number of input variables that don't look like the original input variables.

Feature construction

Feature construction is a commonly used type of feature engineering, and people usually refer to it when they talk about feature engineering. This technique is the process of handcrafting or constructing new features from raw data. In this type of feature engineering, domain knowledge is very useful to manually make up other features from existing ones. Like other feature engineering techniques, the purpose of feature construction is to increase the predictivity of your model. A simple example of feature construction is using the date stamp feature to generate two new features, such as AM and PM, which might be useful to distinguish between day and night. We can also transform/convert noisy numerical features into simpler, nominal ones by calculating the mean value of the noisy feature and then determining whether a given row is more than or less than that mean value.

Titanic example revisited

In this section, we are going to go through the Titanic example again but from a different perspective while using the feature engineering tool. In case you skipped Chapter 2, *Data Modeling in Action - The Titanic Example*, the Titanic example is a Kaggle competition with the purpose of predicting weather a specific passenger survived or not.

During this revisit of the Titanic example, we are going to use the scikit-learn and pandas libraries. So first off, let's start by reading the train and test sets and get some statistics about the data:

```
# reading the train and test sets using pandas
train_data = pd.read_csv('data/train.csv', header=0)
test_data = pd.read_csv('data/test.csv', header=0)

# concatenate the train and test set together for doing the overall feature
engineering stuff
df_titanic_data = pd.concat([train_data, test_data])

# removing duplicate indices due to coming the train and test set by re-
indexing the data
df_titanic_data.reset_index(inplace=True)

# removing the index column the reset_index() function generates
df_titanic_data.drop('index', axis=1, inplace=True)

# index the columns to be 1-based index
df_titanic_data = df_titanic_data.reindex_axis(train_data.columns, axis=1)
```

We need to point out a few things about the preceding code snippet:

- As shown, we have used the `concat` function of pandas to combine the data frames of the train and test sets. This is useful for the feature engineering task as we need a full view of the distribution of the input variables/features.
- After combining both data frames, we need to do some modifications to the output data frame.

Missing values

This step will be the first thing to think of after getting a new dataset from the customer, because there will be missing/incorrect data in nearly every dataset. In the next chapters, you will see that some learning algorithms are able to deal with missing values and others need you to handle missing data. During this example, we are going to use the random forest classifier from scikit-learn, which requires separate handling of missing data.

There are different approaches that you can use to handle missing data.

Removing any sample with missing values in it

This approach won't be a good choice if you have a small dataset with lots of missing values, as removing the samples with missing values will produce useless data. It could be a quick and easy choice if you have lots of data, and removing it won't affect the original dataset much.

Missing value inputting

This approach is useful when you have categorical data. The intuition behind this approach is that missing values may correlate with other variables, and removing them will result in a loss of information that can affect the model significantly.

For example, if we have a binary variable with two possible values, -1 and 1, we can add another value (0) to indicate a missing value. You can use the following code to replace the null values of the **Cabin** feature with U0:

```
# replacing the missing value in cabin variable "U0"
df_titanic_data['Cabin'][df_titanic_data.Cabin.isnull()] = 'U0'
```

Assigning an average value

This is also one of the common approaches because of its simplicity. In the case of a numerical feature, you can just replace the missing values with the mean or median. You can also use this approach in the case of categorical variables by assigning the mode (the value that has the highest occurrence) to the missing values.

The following code assigns the median of the non-missing values of the Fare feature to the missing values:

```
# handling the missing values by replacing it with the median fare
df_titanic_data['Fare'][np.isnan(df_titanic_data['Fare'])] =
df_titanic_data['Fare'].median()
```

Or, you can use the following code to find the value that has the highest occurrence in the Embarked feature and assign it to the missing values:

```
# replacing the missing values with the most common value in the variable
df_titanic_data.Embarked[df_titanic_data.Embarked.isnull()] =
df_titanic_data.Embarked.dropna().mode().values
```

Using a regression or another simple model to predict the values of missing variables

This is the approach that we will use for the Age feature of the Titanic example. The Age feature is an important step towards predicting the survival of passengers, and applying the previous approach by taking the mean will make us lose some information.

In order to predict the missing values, you need to use a supervised learning algorithm that takes the available features as input and the available values of the feature that you want to predict for its missing value as output. In the following code snippet, we are using the random forest classifier to predict the missing values of the Age feature:

```
# Define a helper function that can use RandomForestClassifier for handling
the missing values of the age variable
def set_missing_ages():
    global df_titanic_data

    age_data = df_titanic_data[
        ['Age', 'Embarked', 'Fare', 'Parch', 'SibSp', 'Title_id', 'Pclass',
'Names', 'CabinLetter']]
    input_values_RF =
age_data.loc[(df_titanic_data.Age.notnull())].values[:, 1::]
    target_values_RF =
age_data.loc[(df_titanic_data.Age.notnull())].values[:, 0]

    # Creating an object from the random forest regression function of
sklearn<use the documentation for more details>
    regressor = RandomForestRegressor(n_estimators=2000, n_jobs=-1)

    # building the model based on the input values and target values above
    regressor.fit(input_values_RF, target_values_RF)

    # using the trained model to predict the missing values
    predicted_ages =
regressor.predict(age_data.loc[(df_titanic_data.Age.isnull())].values[:,
1::])
```

```
# Filling the predicted ages in the original titanic dataframe
age_data.loc[(age_data.Age.isnull()), 'Age'] = predicted_ages
```

Feature transformations

In the previous two sections, we covered reading the train and test sets and combining them. We also handled some missing values. Now, we will use the random forest classifier of scikit-learn to predict the survival of passengers. Different implementations of the random forest algorithm accept different types of data. The scikit-learn implementation of random forest accepts only numeric data. So, we need to transform the categorical features into numerical ones.

There are two types of features:

- **Quantitative**: Quantitative features are measured in a numerical scale and can be meaningfully sorted. In the Titanic data samples, the Age feature is an example of a quantitative feature.

- **Qualitative**: Qualitative variables, also called **categorical variables**, are variables that are not numerical. They describe data that fits into categories. In the Titanic data samples, the Embarked (indicates the name of the departure port) feature is an example of a qualitative feature.

We can apply different kinds of transformations to different variables. The following are some approaches that one can use to transform qualitative/categorical features.

Dummy features

These variables are also known as categorical or binary features. This approach will be a good choice if we have a small number of distinct values for the feature to be transformed. In the Titanic data samples, the Embarked feature has only three distinct values (S, C, and Q) that occur frequently. So, we can transform the Embarked feature into three dummy variables, ('Embarked_S', 'Embarked_C', and 'Embarked_Q') to be able to use the random forest classifier.

The following code will show you how to do this kind of transformation:

```
# constructing binary features
def process_embarked():
    global df_titanic_data

    # replacing the missing values with the most common value in the
```

```
variable
    df_titanic_data.Embarked[df.Embarked.isnull()] =
df_titanic_data.Embarked.dropna().mode().values

    # converting the values into numbers
    df_titanic_data['Embarked'] =
pd.factorize(df_titanic_data['Embarked'])[0]

    # binarizing the constructed features
    if keep_binary:
        df_titanic_data = pd.concat([df_titanic_data,
pd.get_dummies(df_titanic_data['Embarked']).rename(
            columns=lambda x: 'Embarked_' + str(x))], axis=1)
```

Factorizing

This approach is used to create a numerical categorical feature from any other feature. In pandas, the `factorize()` function does that. This type of transformation is useful if your feature is an alphanumeric categorical variable. In the Titanic data samples, we can transform the `Cabin` feature into a categorical feature, representing the letter of the cabin:

```
# the cabin number is a sequence of of alphanumerical digits, so we are
going to create some features
# from the alphabetical part of it
df_titanic_data['CabinLetter'] = df_titanic_data['Cabin'].map(lambda l:
get_cabin_letter(l))
df_titanic_data['CabinLetter'] =
pd.factorize(df_titanic_data['CabinLetter'])[0]

def get_cabin_letter(cabin_value):
    # searching for the letters in the cabin alphanumerical value
    letter_match = re.compile("([a-zA-Z]+)").search(cabin_value)

    if letter_match:
        return letter_match.group()
    else:
        return 'U'
```

We can also apply transformations to quantitative features by using one of the following approaches.

Scaling

This kind of transformation can be applied to numerical features only.

For example, in the Titanic data, the `Age` feature can reach 100, but the household income may be in millions. Some models are sensitive to the magnitude of values, so scaling such features will help those models perform better. Also, scaling can be used to squash a variable's values to be within a specific range.

The following code will scale the `Age` feature by removing its mean from each value and scale to the unit variance:

```
# scale by subtracting the mean from each value

scaler_processing = preprocessing.StandardScaler()

df_titanic_data['Age_scaled'] =
scaler_processing.fit_transform(df_titanic_data['Age'])
```

Binning

This kind of quantitative transformation is used to create quantiles. In this case, the quantitative feature values will be the transformed ordered variable. This approach is not a good choice for linear regression, but it might work well for learning algorithms that respond effectively when using ordered/categorical variables.

The following code applies this kind of transformation to the `Fare` feature:

```
# Binarizing the features by binning them into quantiles
df_titanic_data['Fare_bin'] = pd.qcut(df_titanic_data['Fare'], 4)

if keep_binary:
    df_titanic_data = pd.concat(
        [df_titanic_data,
pd.get_dummies(df_titanic_data['Fare_bin']).rename(columns=lambda x:
'Fare_' + str(x))],
        axis=1)
```

Derived features

In the previous section, we applied some transformations to the Titanic data in order to be able to use the random forest classifier of scikit-learn (which only accepts numerical data). In this section, we are going to define another type of variable, which is derived from one or more other features.

Under this definition, we can say that some of the transformations in the previous section are also called **derived features**. In this section, we will look into other, complex transformations.

In the previous sections, we mentioned that you need to use your feature engineering skills to derive new features to enhance the model's predictive power. We have also talked about the importance of feature engineering in the data science pipeline and why you should spend most of your time and effort coming up with useful features. Domain knowledge will be very helpful in this section.

Very simple examples of derived features will be something like extracting the country code and/or region code from a telephone number. You can also extract the country/region from the GPS coordinates.

The Titanic data is a very simple one and doesn't contain a lot of variables to work with, but we can try to derive some features from the text feature that we have in it.

Name

The `name` variable by itself is useless for most datasets, but it has two useful properties. The first one is the length of your name. For example, the length of your name may reflect something about your status and hence your ability to get on a lifeboat:

```python
# getting the different names in the names variable
df_titanic_data['Names'] = df_titanic_data['Name'].map(lambda y:
len(re.split(' ', y)))
```

The second interesting property is the `Name` title, which can also be used to indicate status and/or gender:

```python
# Getting titles for each person
df_titanic_data['Title'] = df_titanic_data['Name'].map(lambda y:
re.compile(", (.*?)\.").findall(y)[0])

# handling the low occurring titles
df_titanic_data['Title'][df_titanic_data.Title == 'Jonkheer'] = 'Master'
df_titanic_data['Title'][df_titanic_data.Title.isin(['Ms', 'Mlle'])] =
'Miss'
df_titanic_data['Title'][df_titanic_data.Title == 'Mme'] = 'Mrs'
df_titanic_data['Title'][df_titanic_data.Title.isin(['Capt', 'Don',
'Major', 'Col', 'Sir'])] = 'Sir'
df_titanic_data['Title'][df_titanic_data.Title.isin(['Dona', 'Lady', 'the
Countess'])] = 'Lady'

# binarizing all the features
```

```
if keep_binary:
    df_titanic_data = pd.concat(
        [df_titanic_data,
pd.get_dummies(df_titanic_data['Title']).rename(columns=lambda x: 'Title_'
+ str(x))],
        axis=1)
```

You can also try to come up with other interesting features from the Name feature. For example, you might think of using the last name feature to find out the size of family members on the Titanic ship.

Cabin

In the Titanic data, the Cabin feature is represented by a letter, which indicates the deck, and a number, which indicates the room number. The room number increases towards the back of the boat, and this will provide some useful measure of the passenger's location. We can also get the status of the passenger from the different decks, and this will help to determine who gets on the lifeboats:

```
# repllacing the missing value in cabin variable "U0"
df_titanic_data['Cabin'][df_titanic_data.Cabin.isnull()] = 'U0'

# the cabin number is a sequence of of alphanumerical digits, so we are
going to create some features
# from the alphabetical part of it
df_titanic_data['CabinLetter'] = df_titanic_data['Cabin'].map(lambda l:
get_cabin_letter(l))
df_titanic_data['CabinLetter'] =
pd.factorize(df_titanic_data['CabinLetter'])[0]

# binarizing the cabin letters features
if keep_binary:
    cletters =
pd.get_dummies(df_titanic_data['CabinLetter']).rename(columns=lambda x:
'CabinLetter_' + str(x))
    df_titanic_data = pd.concat([df_titanic_data, cletters], axis=1)

# creating features from the numerical side of the cabin
df_titanic_data['CabinNumber'] = df_titanic_data['Cabin'].map(lambda x:
get_cabin_num(x)).astype(int) + 1
```

Ticket

The code of the Ticket feature is not immediately clear, but we can do some guesses and try to group them. After looking at the Ticket feature, you may get these clues:

- Almost a quarter of the tickets begin with a character while the rest consist of only numbers.
- The number part of the ticket code seems to have some indications about the class of the passenger. For example, numbers starting with 1 are usually first class tickets, 2 are usually second, and 3 are third. I say *usually* because it holds for the majority of examples, but not all. There are also ticket numbers starting with 4-9, and those are rare and almost exclusively third class.
- Several people can share a ticket number, which might indicate a family or close friends traveling together and acting like a family.

The following code tries to analyze the ticket feature code to come up with preceding clues:

```
# Helper function for constructing features from the ticket variable
def process_ticket():
    global df_titanic_data

    df_titanic_data['TicketPrefix'] = df_titanic_data['Ticket'].map(lambda
y: get_ticket_prefix(y.upper()))
    df_titanic_data['TicketPrefix'] =
df_titanic_data['TicketPrefix'].map(lambda y: re.sub('[\.?\/?]', '', y))
    df_titanic_data['TicketPrefix'] =
df_titanic_data['TicketPrefix'].map(lambda y: re.sub('STON', 'SOTON', y))

    df_titanic_data['TicketPrefixId'] =
pd.factorize(df_titanic_data['TicketPrefix'])[0]

    # binarzing features for each ticket layer
    if keep_binary:
        prefixes =
pd.get_dummies(df_titanic_data['TicketPrefix']).rename(columns=lambda y:
'TicketPrefix_' + str(y))
        df_titanic_data = pd.concat([df_titanic_data, prefixes], axis=1)

    df_titanic_data.drop(['TicketPrefix'], axis=1, inplace=True)

    df_titanic_data['TicketNumber'] = df_titanic_data['Ticket'].map(lambda
y: get_ticket_num(y))
    df_titanic_data['TicketNumberDigits'] =
df_titanic_data['TicketNumber'].map(lambda y: len(y)).astype(np.int)
    df_titanic_data['TicketNumberStart'] =
df_titanic_data['TicketNumber'].map(lambda y: y[0:1]).astype(np.int)
```

```
        df_titanic_data['TicketNumber'] =
df_titanic_data.TicketNumber.astype(np.int)

    if keep_scaled:
        scaler_processing = preprocessing.StandardScaler()
        df_titanic_data['TicketNumber_scaled'] =
scaler_processing.fit_transform(
            df_titanic_data.TicketNumber.reshape(-1, 1))


def get_ticket_prefix(ticket_value):
    # searching for the letters in the ticket alphanumerical value
    match_letter = re.compile("([a-zA-Z\.\/]+)").search(ticket_value)
    if match_letter:
        return match_letter.group()
    else:
        return 'U'


def get_ticket_num(ticket_value):
    # searching for the numbers in the ticket alphanumerical value
    match_number = re.compile("([\d]+$)").search(ticket_value)
    if match_number:
        return match_number.group()
    else:
        return '0'
```

Interaction features

Interaction features are obtained by performing mathematical operations on sets of features and indicate the effect of the relationship between variables. We use basic mathematical operations on the numerical features and see the effects of the relationship between variables:

```
# Constructing features manually based on  the interaction between the
individual features
numeric_features = df_titanic_data.loc[:,
                    ['Age_scaled', 'Fare_scaled', 'Pclass_scaled',
'Parch_scaled', 'SibSp_scaled',
                     'Names_scaled', 'CabinNumber_scaled',
'Age_bin_id_scaled', 'Fare_bin_id_scaled']]
print("\nUsing only numeric features for automated feature generation:\n",
numeric_features.head(10))

new_fields_count = 0
for i in range(0, numeric_features.columns.size - 1):
```

```
        for j in range(0, numeric_features.columns.size - 1):
            if i <= j:
                name = str(numeric_features.columns.values[i]) + "*" +
str(numeric_features.columns.values[j])
                df_titanic_data = pd.concat(
                    [df_titanic_data, pd.Series(numeric_features.iloc[:, i] *
numeric_features.iloc[:, j], name=name)],
                    axis=1)
                new_fields_count += 1
            if i < j:
                name = str(numeric_features.columns.values[i]) + "+" +
str(numeric_features.columns.values[j])
                df_titanic_data = pd.concat(
                    [df_titanic_data, pd.Series(numeric_features.iloc[:, i] +
numeric_features.iloc[:, j], name=name)],
                    axis=1)
                new_fields_count += 1
            if not i == j:
                name = str(numeric_features.columns.values[i]) + "/" +
str(numeric_features.columns.values[j])
                df_titanic_data = pd.concat(
                    [df_titanic_data, pd.Series(numeric_features.iloc[:, i] /
numeric_features.iloc[:, j], name=name)],
                    axis=1)
                name = str(numeric_features.columns.values[i]) + "-" +
str(numeric_features.columns.values[j])
                df_titanic_data = pd.concat(
                    [df_titanic_data, pd.Series(numeric_features.iloc[:, i] -
numeric_features.iloc[:, j], name=name)],
                    axis=1)
                new_fields_count += 2

print("\n", new_fields_count, "new features constructed")
```

This kind of feature engineering can produce lots of features. In the preceding code snippet, we used 9 features to generate 176 interaction features.

We can also remove highly correlated features as the existence of these features won't add any information to the model. We can use Spearman's correlation to identify and remove highly correlated features. The Spearman method has a rank coefficient in its output that can be used to identity the highly correlated features:

```
# using Spearman correlation method to remove the feature that have high
correlation

# calculating the correlation matrix
df_titanic_data_cor = df_titanic_data.drop(['Survived', 'PassengerId'],
```

```
axis=1).corr(method='spearman')

# creating a mask that will ignore correlated ones
mask_ignore = np.ones(df_titanic_data_cor.columns.size) -
np.eye(df_titanic_data_cor.columns.size)
df_titanic_data_cor = mask_ignore * df_titanic_data_cor

features_to_drop = []

# dropping the correclated features
for column in df_titanic_data_cor.columns.values:

    # check if we already decided to drop this variable
    if np.in1d([column], features_to_drop):
        continue

    # finding highly correlacted variables
    corr_vars = df_titanic_data_cor[abs(df_titanic_data_cor[column]) >
0.98].index
    features_to_drop = np.union1d(features_to_drop, corr_vars)

print("\nWe are going to drop", features_to_drop.shape[0], " which are
highly correlated features...\n")
df_titanic_data.drop(features_to_drop, axis=1, inplace=True)
```

The curse of dimensionality

In order to better explain the curse of dimensionality and the problem of overfitting, we are going to go through an example in which we have a set of images. Each image has a cat or a dog in it. So, we would like to build a model that can distinguish between the images with cats and the ones with dogs. Like the fish recognition system in Chapter 1, *Data science - Bird's-eye view*, we need to find an explanatory feature that the learning algorithm can use to distinguish between the two classes (cats and dogs). In this example, we can argue that color is a good descriptor to be used to differentiate between cats and dogs. So the average red, average blue, and average green colors can be used as explanatory features to distinguish between the two classes.

The algorithm will then combine these three features in some way to form a decision boundary between the two classes.

A simple linear combination of the three features can be something like the following:

```
If 0.5*red + 0.3*green + 0.2*blue > 0.6 : return cat;

else return dog;
```

These descriptive features will not be enough to get a good performing classifie, so we can decide to add more features that will enhance the model predictivity to discriminate between cats and dogs. For example, we can consider adding some features such as the texture of the image by calculating the average edge or gradient intensity in both dimensions of the image, X and Y. After adding these two features, the model accuracy will improve. We can even make the model/classifier get more accurate classification power by adding more and more features that are based on color, texture histograms, statistical moments, and so on. We can easily add a few hundred of these features to enhance the model's predictivity. But the counter-intuitive results will be worse after increasing the features beyond some limit. You'll better understand this by looking at *Figure 1*:

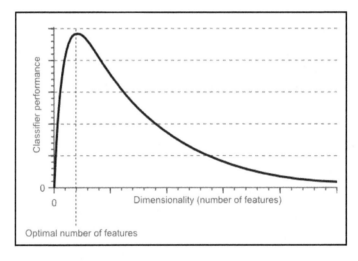

Figure 1: Model performance versus number of features

Figure 1 shows that as the number of features increases, the classifier's performance increases as well, until we reach the optimal number of features. Adding more features based on the same size of the training set will then degrade the classifier's performance.

Avoiding the curse of dimensionality

In the previous sections, we showed that the classifier's performance will decrease when the number of features exceeds a certain optimal point. In theory, if you have infinite training samples, the curse of dimensionality won't exist. So, the optimal number of features is totally dependent on the size of your data.

An approach that will help you to avoid the harm of this curse is to subset M features from the large number of features N, where $M \ll N$. Each feature from M can be a combination of some features in N. There are some algorithms that can do this for you. These algorithms somehow try to find useful, uncorrelated, and linear combinations of the original N features. A commonly used technique for this is **principle component analysis (PCA)**. PCA tries to find a smaller number of features that capture the largest variance of the original data. You can find more insights and a full explanation of PCA at this interesting blog: http://www.visiondummy.com/2014/05/feature-extraction-using-pca/.

A useful and easy way to apply PCA over your original training features is by using the following code:

```
# minimum variance percentage that should be covered by the reduced number
of variables
variance_percentage = .99

# creating PCA object
pca_object = PCA(n_components=variance_percentage)

# trasforming the features
input_values_transformed = pca_object.fit_transform(input_values,
target_values)

# creating a datafram for the transformed variables from PCA
pca_df = pd.DataFrame(input_values_transformed)

print(pca_df.shape[1], " reduced components which describe ",
str(variance_percentage)[1:], "% of the variance")
```

In the Titanic example, we tried to build the classifier with and without applying PCA on the original features. Because we used the random forest classifier at the end, we found that applying PCA isn't very helpful; random forest works very well without any feature transformations, and even correlated features don't really affect the model much.

Titanic example revisited – all together

In this section, we are going to put all the bits and pieces of feature engineering and dimensionality reduction together:

```
import re
import numpy as np
import pandas as pd
import random as rd
from sklearn import preprocessing
from sklearn.cluster import KMeans
from sklearn.ensemble import RandomForestRegressor
from sklearn.decomposition import PCA

# Print options
np.set_printoptions(precision=4, threshold=10000, linewidth=160,
edgeitems=999, suppress=True)
pd.set_option('display.max_columns', None)
pd.set_option('display.max_rows', None)
pd.set_option('display.width', 160)
pd.set_option('expand_frame_repr', False)
pd.set_option('precision', 4)


# constructing binary features
def process_embarked():
    global df_titanic_data

    # replacing the missing values with the most common value in the
variable
    df_titanic_data.Embarked[df.Embarked.isnull()] =
df_titanic_data.Embarked.dropna().mode().values

    # converting the values into numbers
    df_titanic_data['Embarked'] =
pd.factorize(df_titanic_data['Embarked'])[0]

    # binarizing the constructed features
    if keep_binary:
        df_titanic_data = pd.concat([df_titanic_data,
pd.get_dummies(df_titanic_data['Embarked']).rename(
            columns=lambda x: 'Embarked_' + str(x))], axis=1)



# Define a helper function that can use RandomForestClassifier for handling
the missing values of the age variable
```

```python
def set_missing_ages():
    global df_titanic_data

    age_data = df_titanic_data[
        ['Age', 'Embarked', 'Fare', 'Parch', 'SibSp', 'Title_id', 'Pclass',
'Names', 'CabinLetter']]
    input_values_RF =
age_data.loc[(df_titanic_data.Age.notnull())].values[:, 1::]
    target_values_RF =
age_data.loc[(df_titanic_data.Age.notnull())].values[:, 0]

    # Creating an object from the random forest regression function of
sklearn<use the documentation for more details>
    regressor = RandomForestRegressor(n_estimators=2000, n_jobs=-1)

    # building the model based on the input values and target values above
    regressor.fit(input_values_RF, target_values_RF)

    # using the trained model to predict the missing values
    predicted_ages =
regressor.predict(age_data.loc[(df_titanic_data.Age.isnull())].values[:,
1::])

    # Filling the predicted ages in the original titanic dataframe
    age_data.loc[(age_data.Age.isnull()), 'Age'] = predicted_ages


# Helper function for constructing features from the age variable
def process_age():
    global df_titanic_data

    # calling the set_missing_ages helper function to use random forest
regression for predicting missing values of age
    set_missing_ages()

    # # scale the age variable by centering it around the mean with a unit
variance
    # if keep_scaled:
    # scaler_preprocessing = preprocessing.StandardScaler()
    # df_titanic_data['Age_scaled'] =
scaler_preprocessing.fit_transform(df_titanic_data.Age.reshape(-1, 1))

    # construct a feature for children
    df_titanic_data['isChild'] = np.where(df_titanic_data.Age < 13, 1, 0)

    # bin into quartiles and create binary features
    df_titanic_data['Age_bin'] = pd.qcut(df_titanic_data['Age'], 4)
```

```python
    if keep_binary:
        df_titanic_data = pd.concat(
            [df_titanic_data,
pd.get_dummies(df_titanic_data['Age_bin']).rename(columns=lambda y: 'Age_'
+ str(y))],
            axis=1)

    if keep_bins:
        df_titanic_data['Age_bin_id'] =
pd.factorize(df_titanic_data['Age_bin'])[0] + 1

    if keep_bins and keep_scaled:
        scaler_processing = preprocessing.StandardScaler()
        df_titanic_data['Age_bin_id_scaled'] =
scaler_processing.fit_transform(
            df_titanic_data.Age_bin_id.reshape(-1, 1))

    if not keep_strings:
        df_titanic_data.drop('Age_bin', axis=1, inplace=True)


# Helper function for constructing features from the passengers/crew names
def process_name():
    global df_titanic_data

    # getting the different names in the names variable
    df_titanic_data['Names'] = df_titanic_data['Name'].map(lambda y:
len(re.split(' ', y)))

    # Getting titles for each person
    df_titanic_data['Title'] = df_titanic_data['Name'].map(lambda y:
re.compile(", (.*?)\.").findall(y)[0])

    # handling the low occurring titles
    df_titanic_data['Title'][df_titanic_data.Title == 'Jonkheer'] =
'Master'
    df_titanic_data['Title'][df_titanic_data.Title.isin(['Ms', 'Mlle'])] =
'Miss'
    df_titanic_data['Title'][df_titanic_data.Title == 'Mme'] = 'Mrs'
    df_titanic_data['Title'][df_titanic_data.Title.isin(['Capt', 'Don',
'Major', 'Col', 'Sir'])] = 'Sir'
    df_titanic_data['Title'][df_titanic_data.Title.isin(['Dona', 'Lady',
'the Countess'])] = 'Lady'

    # binarizing all the features
    if keep_binary:
        df_titanic_data = pd.concat(
            [df_titanic_data,
```

```
pd.get_dummies(df_titanic_data['Title']).rename(columns=lambda x: 'Title_'
+ str(x))],
            axis=1)

    # scaling
    if keep_scaled:
        scaler_preprocessing = preprocessing.StandardScaler()
        df_titanic_data['Names_scaled'] =
scaler_preprocessing.fit_transform(df_titanic_data.Names.reshape(-1, 1))

    # binning
    if keep_bins:
        df_titanic_data['Title_id'] =
pd.factorize(df_titanic_data['Title'])[0] + 1

    if keep_bins and keep_scaled:
        scaler = preprocessing.StandardScaler()
        df_titanic_data['Title_id_scaled'] =
scaler.fit_transform(df_titanic_data.Title_id.reshape(-1, 1))



# Generate features from the cabin input variable
def process_cabin():
    # refering to the global variable that contains the titanic examples
    global df_titanic_data

    # repllacing the missing value in cabin variable "U0"
    df_titanic_data['Cabin'][df_titanic_data.Cabin.isnull()] = 'U0'

    # the cabin number is a sequence of of alphanumerical digits, so we are
going to create some features
    # from the alphabetical part of it
    df_titanic_data['CabinLetter'] = df_titanic_data['Cabin'].map(lambda l:
get_cabin_letter(l))
    df_titanic_data['CabinLetter'] =
pd.factorize(df_titanic_data['CabinLetter'])[0]

    # binarizing the cabin letters features
    if keep_binary:
        cletters =
pd.get_dummies(df_titanic_data['CabinLetter']).rename(columns=lambda x:
'CabinLetter_' + str(x))
        df_titanic_data = pd.concat([df_titanic_data, cletters], axis=1)

    # creating features from the numerical side of the cabin
    df_titanic_data['CabinNumber'] = df_titanic_data['Cabin'].map(lambda x:
get_cabin_num(x)).astype(int) + 1
```

```
    # scaling the feature
    if keep_scaled:
        scaler_processing = preprocessing.StandardScaler() # handling the
missing values by replacing it with the median feare
    df_titanic_data['Fare'][np.isnan(df_titanic_data['Fare'])] =
df_titanic_data['Fare'].median()
    df_titanic_data['CabinNumber_scaled'] =
scaler_processing.fit_transform(df_titanic_data.CabinNumber.reshape(-1, 1))


def get_cabin_letter(cabin_value):
    # searching for the letters in the cabin alphanumerical value
    letter_match = re.compile("([a-zA-Z]+)").search(cabin_value)

    if letter_match:
        return letter_match.group()
    else:
        return 'U'


def get_cabin_num(cabin_value):
    # searching for the numbers in the cabin alphanumerical value
    number_match = re.compile("([0-9]+)").search(cabin_value)

    if number_match:
        return number_match.group()
    else:
        return 0



# helper function for constructing features from the ticket fare variable
def process_fare():
    global df_titanic_data

    # handling the missing values by replacing it with the median feare
    df_titanic_data['Fare'][np.isnan(df_titanic_data['Fare'])] =
df_titanic_data['Fare'].median()

    # zeros in the fare will cause some division problems so we are going
to set them to 1/10th of the lowest fare
    df_titanic_data['Fare'][np.where(df_titanic_data['Fare'] == 0)[0]] =
df_titanic_data['Fare'][
df_titanic_data['Fare'].nonzero()[
0]].min() / 10

    # Binarizing the features by binning them into quantiles
    df_titanic_data['Fare_bin'] = pd.qcut(df_titanic_data['Fare'], 4)
```

```
    if keep_binary:
        df_titanic_data = pd.concat(
            [df_titanic_data,
pd.get_dummies(df_titanic_data['Fare_bin']).rename(columns=lambda x:
'Fare_' + str(x))],
            axis=1)

    # binning
    if keep_bins:
        df_titanic_data['Fare_bin_id'] =
pd.factorize(df_titanic_data['Fare_bin'])[0] + 1

    # scaling the value
    if keep_scaled:
        scaler_processing = preprocessing.StandardScaler()
        df_titanic_data['Fare_scaled'] =
scaler_processing.fit_transform(df_titanic_data.Fare.reshape(-1, 1))

    if keep_bins and keep_scaled:
        scaler_processing = preprocessing.StandardScaler()
        df_titanic_data['Fare_bin_id_scaled'] =
scaler_processing.fit_transform(
            df_titanic_data.Fare_bin_id.reshape(-1, 1))

    if not keep_strings:
        df_titanic_data.drop('Fare_bin', axis=1, inplace=True)


# Helper function for constructing features from the ticket variable
def process_ticket():
    global df_titanic_data

    df_titanic_data['TicketPrefix'] = df_titanic_data['Ticket'].map(lambda
y: get_ticket_prefix(y.upper()))
    df_titanic_data['TicketPrefix'] =
df_titanic_data['TicketPrefix'].map(lambda y: re.sub('[\.?\/?]', '', y))
    df_titanic_data['TicketPrefix'] =
df_titanic_data['TicketPrefix'].map(lambda y: re.sub('STON', 'SOTON', y))

    df_titanic_data['TicketPrefixId'] =
pd.factorize(df_titanic_data['TicketPrefix'])[0]

    # binarzing features for each ticket layer
    if keep_binary:
        prefixes =
pd.get_dummies(df_titanic_data['TicketPrefix']).rename(columns=lambda y:
'TicketPrefix_' + str(y))
        df_titanic_data = pd.concat([df_titanic_data, prefixes], axis=1)
```

```
    df_titanic_data.drop(['TicketPrefix'], axis=1, inplace=True)

    df_titanic_data['TicketNumber'] = df_titanic_data['Ticket'].map(lambda
y: get_ticket_num(y))
    df_titanic_data['TicketNumberDigits'] =
df_titanic_data['TicketNumber'].map(lambda y: len(y)).astype(np.int)
    df_titanic_data['TicketNumberStart'] =
df_titanic_data['TicketNumber'].map(lambda y: y[0:1]).astype(np.int)

    df_titanic_data['TicketNumber'] =
df_titanic_data.TicketNumber.astype(np.int)

    if keep_scaled:
        scaler_processing = preprocessing.StandardScaler()
        df_titanic_data['TicketNumber_scaled'] =
scaler_processing.fit_transform(
            df_titanic_data.TicketNumber.reshape(-1, 1))


def get_ticket_prefix(ticket_value):
    # searching for the letters in the ticket alphanumerical value
    match_letter = re.compile("([a-zA-Z\.\/]+)").search(ticket_value)
    if match_letter:
        return match_letter.group()
    else:
        return 'U'


def get_ticket_num(ticket_value):
    # searching for the numbers in the ticket alphanumerical value
    match_number = re.compile("([\d]+$)").search(ticket_value)
    if match_number:
        return match_number.group()
    else:
        return '0'


# constructing features from the passenger class variable
def process_PClass():
    global df_titanic_data

    # using the most frequent value(mode) to replace the messing value
    df_titanic_data.Pclass[df_titanic_data.Pclass.isnull()] =
df_titanic_data.Pclass.dropna().mode().values

    # binarizing the features
    if keep_binary:
        df_titanic_data = pd.concat(
```

```
            [df_titanic_data,
pd.get_dummies(df_titanic_data['Pclass']).rename(columns=lambda y:
'Pclass_' + str(y))],
            axis=1)

    if keep_scaled:
        scaler_preprocessing = preprocessing.StandardScaler()
        df_titanic_data['Pclass_scaled'] =
scaler_preprocessing.fit_transform(df_titanic_data.Pclass.reshape(-1, 1))


# constructing features based on the family variables subh as SibSp and
Parch
def process_family():
    global df_titanic_data

    # ensuring that there's no zeros to use interaction variables
    df_titanic_data['SibSp'] = df_titanic_data['SibSp'] + 1
    df_titanic_data['Parch'] = df_titanic_data['Parch'] + 1

    # scaling
    if keep_scaled:
        scaler_preprocessing = preprocessing.StandardScaler()
        df_titanic_data['SibSp_scaled'] =
scaler_preprocessing.fit_transform(df_titanic_data.SibSp.reshape(-1, 1))
        df_titanic_data['Parch_scaled'] =
scaler_preprocessing.fit_transform(df_titanic_data.Parch.reshape(-1, 1))

    # binarizing all the features
    if keep_binary:
        sibsps_var =
pd.get_dummies(df_titanic_data['SibSp']).rename(columns=lambda y: 'SibSp_'
+ str(y))
        parchs_var =
pd.get_dummies(df_titanic_data['Parch']).rename(columns=lambda y: 'Parch_'
+ str(y))
        df_titanic_data = pd.concat([df_titanic_data, sibsps_var,
parchs_var], axis=1)


# binarzing the sex variable
def process_sex():
    global df_titanic_data
    df_titanic_data['Gender'] = np.where(df_titanic_data['Sex'] == 'male',
1, 0)


# dropping raw original
```

```
def process_drops():
    global df_titanic_data
    drops = ['Name', 'Names', 'Title', 'Sex', 'SibSp', 'Parch', 'Pclass',
'Embarked', \
             'Cabin', 'CabinLetter', 'CabinNumber', 'Age', 'Fare',
'Ticket', 'TicketNumber']
    string_drops = ['Title', 'Name', 'Cabin', 'Ticket', 'Sex', 'Ticket',
'TicketNumber']
    if not keep_raw:
        df_titanic_data.drop(drops, axis=1, inplace=True)
    elif not keep_strings:
        df_titanic_data.drop(string_drops, axis=1, inplace=True)


# handling all the feature engineering tasks
def get_titanic_dataset(binary=False, bins=False, scaled=False,
strings=False, raw=True, pca=False, balanced=False):
    global keep_binary, keep_bins, keep_scaled, keep_raw, keep_strings,
df_titanic_data
    keep_binary = binary
    keep_bins = bins
    keep_scaled = scaled
    keep_raw = raw
    keep_strings = strings

    # reading the train and test sets using pandas
    train_data = pd.read_csv('data/train.csv', header=0)
    test_data = pd.read_csv('data/test.csv', header=0)

    # concatenate the train and test set together for doing the overall
feature engineering stuff
    df_titanic_data = pd.concat([train_data, test_data])

    # removing duplicate indices due to coming the train and test set by
re-indexing the data
    df_titanic_data.reset_index(inplace=True)

    # removing the index column the reset_index() function generates
    df_titanic_data.drop('index', axis=1, inplace=True)

    # index the columns to be 1-based index
    df_titanic_data = df_titanic_data.reindex_axis(train_data.columns,
axis=1)

    # processing the titanic raw variables using the helper functions that
we defined above
    process_cabin()
    process_ticket()
```

```
process_name()
process_fare()
process_embarked()
process_family()
process_sex()
process_PClass()
process_age()
process_drops()

# move the survived column to be the first
columns_list = list(df_titanic_data.columns.values)
columns_list.remove('Survived')
new_col_list = list(['Survived'])
new_col_list.extend(columns_list)
df_titanic_data = df_titanic_data.reindex(columns=new_col_list)

print("Starting with", df_titanic_data.columns.size,
        "manually constructing features based on the interaction between
them...\n", df_titanic_data.columns.values)

# Constructing features manually based on the interaction between the
individual features
numeric_features = df_titanic_data.loc[:,
                    ['Age_scaled', 'Fare_scaled', 'Pclass_scaled',
'Parch_scaled', 'SibSp_scaled',
                        'Names_scaled', 'CabinNumber_scaled',
'Age_bin_id_scaled', 'Fare_bin_id_scaled']]
    print("\nUsing only numeric features for automated feature
generation:\n", numeric_features.head(10))

new_fields_count = 0
for i in range(0, numeric_features.columns.size - 1):
    for j in range(0, numeric_features.columns.size - 1):
        if i <= j:
            name = str(numeric_features.columns.values[i]) + "*" +
str(numeric_features.columns.values[j])
                df_titanic_data = pd.concat(
                    [df_titanic_data, pd.Series(numeric_features.iloc[:, i]
* numeric_features.iloc[:, j], name=name)],
                    axis=1)
                new_fields_count += 1
        if i < j:
            name = str(numeric_features.columns.values[i]) + "+" +
str(numeric_features.columns.values[j])
                df_titanic_data = pd.concat(
                    [df_titanic_data, pd.Series(numeric_features.iloc[:, i]
+ numeric_features.iloc[:, j], name=name)],
                    axis=1)
```

```
                    new_fields_count += 1
              if not i == j:
                    name = str(numeric_features.columns.values[i]) + "/" +
str(numeric_features.columns.values[j])
                        df_titanic_data = pd.concat(
                            [df_titanic_data, pd.Series(numeric_features.iloc[:, i]
/ numeric_features.iloc[:, j], name=name)],
                        axis=1)
                    name = str(numeric_features.columns.values[i]) + "-" +
str(numeric_features.columns.values[j])
                        df_titanic_data = pd.concat(
                            [df_titanic_data, pd.Series(numeric_features.iloc[:, i]
- numeric_features.iloc[:, j], name=name)],
                        axis=1)
                    new_fields_count += 2

    print("\n", new_fields_count, "new features constructed")

    # using Spearman correlation method to remove the feature that have
high correlation

    # calculating the correlation matrix
    df_titanic_data_cor = df_titanic_data.drop(['Survived', 'PassengerId'],
axis=1).corr(method='spearman')

    # creating a mask that will ignore correlated ones
    mask_ignore = np.ones(df_titanic_data_cor.columns.size) -
np.eye(df_titanic_data_cor.columns.size)
    df_titanic_data_cor = mask_ignore * df_titanic_data_cor

    features_to_drop = []

    # dropping the correclated features
    for column in df_titanic_data_cor.columns.values:

        # check if we already decided to drop this variable
        if np.in1d([column], features_to_drop):
            continue

        # finding highly correclacted variables
        corr_vars = df_titanic_data_cor[abs(df_titanic_data_cor[column]) >
0.98].index
        features_to_drop = np.union1d(features_to_drop, corr_vars)

    print("\nWe are going to drop", features_to_drop.shape[0], " which are
highly correlated features...\n")
    df_titanic_data.drop(features_to_drop, axis=1, inplace=True)
```

```
    # splitting the dataset to train and test and do PCA
    train_data = df_titanic_data[:train_data.shape[0]]
    test_data = df_titanic_data[test_data.shape[0]:]

    if pca:
        print("reducing number of variables...")
        train_data, test_data = reduce(train_data, test_data)
    else:
        # drop the empty 'Survived' column for the test set that was
created during set concatenation
        test_data.drop('Survived', axis=1, inplace=True)

    print("\n", train_data.columns.size, "initial features generated...\n")
# , input_df.columns.values

    return train_data, test_data


# reducing the dimensionality for the training and testing set
def reduce(train_data, test_data):
    # join the full data together
    df_titanic_data = pd.concat([train_data, test_data])
    df_titanic_data.reset_index(inplace=True)
    df_titanic_data.drop('index', axis=1, inplace=True)
    df_titanic_data = df_titanic_data.reindex_axis(train_data.columns,
axis=1)

    # converting the survived column to series
    survived_series = pd.Series(df_titanic_data['Survived'],
name='Survived')

    print(df_titanic_data.head())

    # getting the input and target values
    input_values = df_titanic_data.values[:, 1::]
    target_values = df_titanic_data.values[:, 0]

    print(input_values[0:10])

    # minimum variance percentage that should be covered by the reduced
number of variables
    variance_percentage = .99

    # creating PCA object
    pca_object = PCA(n_components=variance_percentage)

    # trasforming the features
    input_values_transformed = pca_object.fit_transform(input_values,
```

```
    target_values)

        # creating a datafram for the transformed variables from PCA
        pca_df = pd.DataFrame(input_values_transformed)

        print(pca_df.shape[1], " reduced components which describe ",
    str(variance_percentage)[1:], "% of the variance")

        # constructing a new dataframe that contains the newly reduced vars of
    PCA
        df_titanic_data = pd.concat([survived_series, pca_df], axis=1)

        # split into separate input and test sets again
        train_data = df_titanic_data[:train_data.shape[0]]
        test_data = df_titanic_data[test_data.shape[0]:]
        test_data.reset_index(inplace=True)
        test_data.drop('index', axis=1, inplace=True)
        test_data.drop('Survived', axis=1, inplace=True)

        return train_data, test_data


    # Calling the helper functions
    if __name__ == '__main__':
        train, test = get_titanic_dataset(bins=True, scaled=True, binary=True)
        initial_drops = ['PassengerId']
        train.drop(initial_drops, axis=1, inplace=True)
        test.drop(initial_drops, axis=1, inplace=True)

        train, test = reduce(train, test)

        print(train.columns.values)
```

Bias-variance decomposition

In the previous section, we knew how to select the best hyperparameters for our model. This set of best hyperparameters was chosen based on the measure of minimizing the cross validated error. Now, we need to see how the model will perform over the unseen data, or the so-called out-of-sample data, which refers to new data samples that haven't been seen during the model training phase.

Consider the following example: we have a data sample of size 10,000, and we are going to train the same model with different train set sizes and plot the test error at each step. For example, we are going to take out 1,000 as a test set and use the other 9,000 for training. So for the first training round, we will randomly select a train set of size 100 out of those 9,000 items. We'll train the model based on the *best* selected set of hyperparameters, test the model with the test set, and finally plot the train (in-sample) error and the test (out-of-sample) error. We repeat this training, testing, and plotting operation for different train sizes (for example, repeat with 500 out of the 9,000, then 1,000 out of the 9,000, and so on).

After doing all this training, testing, and plotting, we will get a graph of two curves, representing the train and test errors with the same model but across different train set sizes. From this graph, we will get to know how good our model is.

The output graph, which will contain two curves representing the training and testing error, will be one of the four possible shapes shown in *Figure 2*. The source of this different shapes is Andrew Ng's Machine Learning course on Coursera (`https://www.coursera.org/learn/machine-learning`). It's a great course with lots of insights and best practices for machine learning newbies:

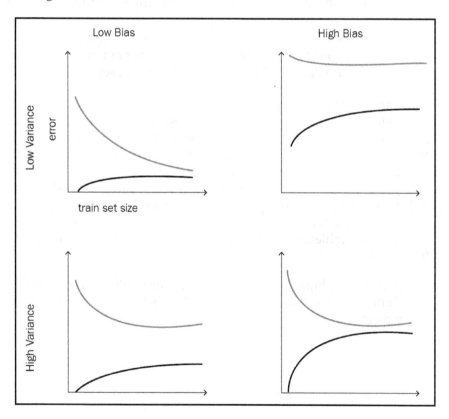

Figure 2: Possible shapes for plotting the training and testing error over different training set sizes

So, when should we accept our model and put it into production? And when do we know that our model is not performing well over the test set and hence won't have a bad generalization error? The answer to these questions depends on the shape that you get from plotting the train error versus the test error on different training set sizes:

- If your shape looks like the *top left* one, it represents a low training error and generalizes well over the test set. This shape is a winner and you should go ahead and use this model in production.
- If your shape is similar to the *top right* one, it represents a high training error (the model didn't manage to learn from the training samples) and even has worse generalization performance over the test set. This shape is a complete failure and you need to go back and see what's wrong with your data, chosen learning algorithm, and/or selected hyperparameters.
- If your shape is similar to the *bottom left* one, it represents a bad training error as the model didn't manage to capture the underlying structure of the data, which also fits the new test data.
- If your shape is similar to the *bottom right* one, it represents high bias and variance. This means that your model hasn't figured out the training data very well and hence didn't generalize well over the testing set.

Bias and variance are the components that we can use to figure out how good our model is. In supervised learning, there are two opposing sources of errors, and using the learning curves in *Figure 2*, we can figure out due to which component(s) our model is suffering. The problem of having high variance and low bias is called **overfitting**, which means that the model performed well over the training samples but didn't generalize well on the test set. On the other hand, the problem of having high bias and low variance is called **underfitting**, which means that the model didn't make use of the data and didn't manage to estimate the output/target from the input features. There are different approaches one can use to avoid getting into one of these problems. But usually, enhancing one of them will come at the expense of the second one.

We can solve the situation of high variance by adding more features from which the model can learn. This solution will most likely increase the bias, so you need to make some sort of trade-off between them.

Learning visibility

There are lots of great data science algorithms that one can use to solve problems in different domains, but the key component that makes the learning process visible is having enough data. You might ask how much data is needed for the learning process to be visible and worth doing. As a rule of thumb, researchers and machine learning practitioners agree that you need to have data samples at least 10 times the number of **degrees of freedom** in your model.

For example, in the case of linear models, the degree of freedom represents the number of features that you have in your dataset. If you have 50 explanatory features in your data, then you need at least 500 data samples/observations in your data.

Breaking the rule of thumb

In practice, you can get away with this rule and do learning with less than 10 times the number of features in your data; this mostly happens if your model is simple and you are using something called **regularization** (addressed in the next chapter).

Jake Vanderplas wrote an article (`https://jakevdp.github.io/blog/2015/07/06/model-complexity-myth/`) to show that one can learn even if the data has more parameters than examples. To demonstrate this, he used regularization.

Summary

In this chapter, we covered the most important tools that machine learning practitioners use in order to make sense of their data and get the learning algorithm to get the most out of their data.

Feature engineering was the first and commonly used tool in data science; it's a must-have component in any data science pipeline. The purpose of this tool is to make better representations for your data and increase the predictive power of your model.

We saw how a large number of features can be problematic and lead to worse classifier performance. We also saw that there is an optimal number of features that should be used to get the maximum model performance, and this optimal number of features is a function of the number of data samples/observations you got.

Subsequently, we introduced one of the most powerful tools, which is bias-variance decomposition. This tool is widely used to test how good the model is over the test set.

Finally, we went through learning visibility, which answers the question of how much data we should need in order to get in business and do machine learning. The rule of thumb showed that we need data samples/observations at least 10 times the number of features in your data. However, this rule of thumb can be broken by using another tool called regularization, which will be addressed in more detail in the next chapter.

Next up, we are going to continue to increase our data science tools that we can use to drive meaningful analytics from our data, and face some daily problems of applying machine learning.

4
Get Up and Running with TensorFlow

In this chapter, we are going to give an overview of one of the most widely used deep learning frameworks. TensorFlow has big community support that is growing day by day, which makes it a good option for building your complex deep learning applications. From the TensorFlow website:

> *"TensorFlow is an open source software library for numerical computation using data flow graphs. Nodes in the graph represent mathematical operations, while the graph edges represent the multidimensional data arrays (tensors) communicated between them. The flexible architecture allows you to deploy computation to one or more CPUs or GPUs in a desktop, server, or mobile device with a single API. TensorFlow was originally developed by researchers and engineers working on the Google Brain Team within Google's Machine Intelligence research organization for the purposes of conducting machine learning and deep neural networks research, but the system is general enough to be applicable in a wide variety of other domains as well."*

The following topics are going to be covered in this chapter:

- TensorFlow installation
- The TensorFlow environment
- Computational graphs
- TensorFlow data types, variables, and placeholders
- Getting output from TensorFlow
- TensorBoard—visualizing learning

TensorFlow installation

TensorFlow installation comes with two modes: CPU and GPU. We will start off the installation tutorial by installing TensorFlow in GPU mode.

TensorFlow GPU installation for Ubuntu 16.04

The GPU mode installation of TensorFlow requires an up-to-date installation of the NVIDIA drivers because the GPU version of TensorFlow only supports CUDA at the moment. The following section will take you through a step-by-step process of installing NVIDIA drivers and CUDA 8.

Installing NVIDIA drivers and CUDA 8

First off, you need to install the correct NVIDIA driver based on your GPU. I have a GeForce GTX 960M GPU, so I will go ahead and install `nvidia-375` (if you have a different GPU, you can use the NVIDIA search tool http://www.nvidia.com/Download/index.aspx to help you find your correct driver version). If you want to know your machine's GPU, you can issue the following command in the terminal:

```
lspci | grep -i nvidia
```

You should get the following output in the terminal:

```
ahmed@ahmed-Inspiron-7559:~$ lspci | grep -i nvidia
02:00.0 3D controller: NVIDIA Corporation GM107M [GeForce GTX 960M] (rev a2)
ahmed@ahmed-Inspiron-7559:~$ 
```

Next, we need to add a proprietary repository of NVIDIA drivers to be able to install the drivers using `apt-get`:

```
sudo add-apt-repository ppa:graphics-drivers/ppa
sudo apt-get update
sudo apt-get install nvidia-375
```

After successfully installing the NVIDIA drivers, restart the machine. To verify whether the drivers installed correctly, issue the following command in the Terminal:

```
cat /proc/driver/nvidia/version
```

You should get the following output in the Terminal:

```
ahmed@ahmed-Inspiron-7559:~$ cat /proc/driver/nvidia/version
NVRM version: NVIDIA UNIX x86_64 Kernel Module  375.26  Thu Dec  8 18:36:43 PST
2016
GCC version:  gcc version 5.4.0 20160609 (Ubuntu 5.4.0-6ubuntu1~16.04.4)
ahmed@ahmed-Inspiron-7559:~$
```

Next, we need to install CUDA 8. Open the following CUDA download link: `https://developer.nvidia.com/cuda-downloads`. Select your operating system, architecture, distribution, version, and finally, installer type as per the following screenshot:

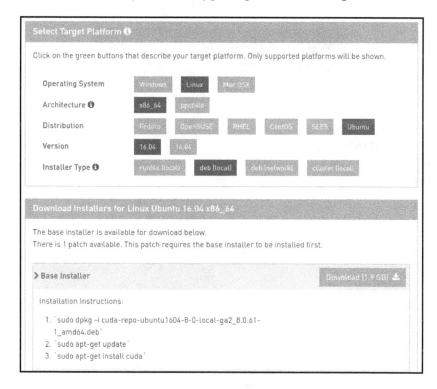

The installer file is about 2 GB. You need to issue the following installation instructions:

```
sudo dpkg -i cuda-repo-ubuntu1604-8-0-local-ga2_8.0.61-1_amd64.deb
sudo apt-get update
sudo apt-get install cuda
```

Next, we need to add the libraries to the `.bashrc` file by issuing the following commands:

```
echo 'export PATH=/usr/local/cuda/bin:$PATH' >> ~/.bashrc

echo 'export LD_LIBRARY_PATH=/usr/local/cuda/lib64:$LD_LIBRARY_PATH' >>
~/.bashrc

source ~/.bashrc
```

Next, you need to verify the installation of CUDA 8 by issuing the following command:

```
nvcc -V
```

You should get the following output in the terminal:

```
ahmed@ahmed-Inspiron-7559:~$ nvcc -V
nvcc: NVIDIA (R) Cuda compiler driver
Copyright (c) 2005-2016 NVIDIA Corporation
Built on Tue_Jan_10_13:22:03_CST_2017
Cuda compilation tools, release 8.0, V8.0.61
ahmed@ahmed-Inspiron-7559:~$
```

Finally, in this section, we need to install cuDNN 6.0. The **NVIDIA CUDA Deep Neural Network library (cuDNN)** is a GPU-accelerated library of primitives for deep neural networks. You can download it from NVIDIA's web page. Issue the following commands to extract and install cuDNN:

```
cd ~/Downloads/

tar xvf cudnn*.tgz

cd cuda

sudo cp */*.h /usr/local/cuda/include/

sudo cp */libcudnn* /usr/local/cuda/lib64/

sudo chmod a+r /usr/local/cuda/lib64/libcudnn*
```

To ensure that your installation has been successful, you can use the `nvidia-smi` tool in the terminal. If you had a successful installation, this tool will provide you with a monitoring information such as RAM and the running process for your GPU.

Installing TensorFlow

After preparing the GPU environment for TensorFlow, we are now ready to install TensorFlow in GPU mode. But for going through the TensorFlow installation process, you can first install a few helpful Python packages that will help you in the next chapters and make your development environment easier.

We can start by installing some data manipulation, analysis, and visualization libraries by issuing the following commands:

```
sudo apt-get update && apt-get install -y python-numpy python-scipy python-
nose python-h5py python-skimage python-matplotlib python-pandas python-
sklearn python-sympy

sudo apt-get clean && sudo apt-get autoremove

sudo rm -rf /var/lib/apt/lists/*
```

Next, you can install more useful libraries, such as the virtual environment, Jupyter Notebook, and so on:

```
sudo apt-get update

sudo apt-get install git python-dev python3-dev python-numpy python3-numpy
build-essential  python-pip python3-pip python-virtualenv swig python-wheel
libcurl3-dev

sudo apt-get install -y libfreetype6-dev libpng12-dev

pip3 install -U matplotlib ipython[all] jupyter pandas scikit-image
```

Finally, we can start to install TensorFlow in GPU mode by issuing the following command:

```
pip3 install --upgrade tensorflow-gpu
```

You can verify the successful installation of TensorFlow using Python:

```
python3
>>> import tensorflow as tf
>>> a = tf.constant(5)
>>> b = tf.constant(6)
>>> sess = tf.Session()
>>> sess.run(a+b)
// this should print bunch of messages showing device status etc. // If
everything goes well, you should see gpu listed in device
>>> sess.close()
```

You should get the following output in the terminal:

```
ahmed@ahmed-Inspiron-7559:~$ python3
Python 3.6.0 |Anaconda 4.3.0 (64-bit)| (default, Dec 23 2016, 12:22:00)
[GCC 4.4.7 20120313 (Red Hat 4.4.7-1)] on linux
Type "help", "copyright", "credits" or "license" for more information.
>>> import tensorflow as tf
I tensorflow/stream_executor/dso_loader.cc:128] successfully opened CUDA library libcublas.so locally
I tensorflow/stream_executor/dso_loader.cc:128] successfully opened CUDA library libcudnn.so locally
I tensorflow/stream_executor/dso_loader.cc:128] successfully opened CUDA library libcufft.so locally
I tensorflow/stream_executor/dso_loader.cc:128] successfully opened CUDA library libcuda.so.1 locally
I tensorflow/stream_executor/dso_loader.cc:128] successfully opened CUDA library libcurand.so locally
>>> a = tf.constant(5)
>>> b = tf.constant(6)
>>> sess = tf.Session()
I tensorflow/stream_executor/cuda/cuda_gpu_executor.cc:937] successful NUMA node read from SysFS had
negative value (-1), but there must be at least one NUMA node, so returning NUMA node zero
I tensorflow/core/common_runtime/gpu/gpu_device.cc:885] Found device 0 with properties:
name: GeForce GTX 960M
major: 5 minor: 0 memoryClockRate (GHz) 1.176
pciBusID 0000:02:00.0
Total memory: 3.95GiB
Free memory: 3.43GiB
I tensorflow/core/common_runtime/gpu/gpu_device.cc:906] DMA: 0
I tensorflow/core/common_runtime/gpu/gpu_device.cc:916] 0:   Y
I tensorflow/core/common_runtime/gpu/gpu_device.cc:975] Creating TensorFlow device (/gpu:0) -> (devic
e: 0, name: GeForce GTX 960M, pci bus id: 0000:02:00.0)
>>> sess.run(a+b)
11
>>> sess.close()
>>>
```

TensorFlow CPU installation for Ubuntu 16.04

In this section, we are going to install the CPU version, which doesn't require any drivers prior to installation. So, let's start off by installing some useful packages for data manipulation and visualization:

```
sudo apt-get update && apt-get install -y python-numpy python-scipy python-
nose python-h5py python-skimage python-matplotlib python-pandas python-
sklearn python-sympy

sudo apt-get clean && sudo apt-get autoremove

sudo rm -rf /var/lib/apt/lists/*
```

Next, you can install more useful libraries, such as the virtual environment, Jupyter Notebook, and so on:

```
sudo apt-get update

sudo apt-get install git python-dev python3-dev python-numpy python3-numpy
build-essential  python-pip python3-pip python-virtualenv swig python-wheel
```

```
libcurl3-dev

sudo apt-get install -y libfreetype6-dev libpng12-dev

pip3 install -U matplotlib ipython[all] jupyter pandas scikit-image
```

Finally, you can install the latest TensorFlow in CPU mode by issuing the following command:

```
pip3 install --upgrade tensorflow
```

You can check whether TensorFlow was installed successfully be running the following TensorFlow statements:

```
python3
>>> import tensorflow as tf
>>> a = tf.constant(5)
>>> b = tf.constant(6)
>>> sess = tf.Session()
>>> sess.run(a+b)
>> sess.close()
```

You should get the following output in the terminal:

```
ahmed@ahmed-Inspiron-7559:~$ python3
Python 3.6.0 |Anaconda 4.3.0 (64-bit)| (default, Dec 23 2016, 12:22:00)
[GCC 4.4.7 20120313 (Red Hat 4.4.7-1)] on linux
Type "help", "copyright", "credits" or "license" for more information.
>>> import tensorflow as tf
>>> a = tf.constant(5)
>>> b = tf.constant(6)
>>> sess = tf.Session()
2017-09-11 15:39:50.792072: W tensorflow/core/platform/cpu_feature_guard.cc:45] The TensorFlow librar
y wasn't compiled to use SSE4.1 instructions, but these are available on your machine and could speed
up CPU computations.
2017-09-11 15:39:50.792125: W tensorflow/core/platform/cpu_feature_guard.cc:45] The TensorFlow librar
y wasn't compiled to use SSE4.2 instructions, but these are available on your machine and could speed
up CPU computations.
2017-09-11 15:39:50.792143: W tensorflow/core/platform/cpu_feature_guard.cc:45] The TensorFlow librar
y wasn't compiled to use AVX instructions, but these are available on your machine and could speed up
 CPU computations.
2017-09-11 15:39:50.792158: W tensorflow/core/platform/cpu_feature_guard.cc:45] The TensorFlow librar
y wasn't compiled to use AVX2 instructions, but these are available on your machine and could speed u
p CPU computations.
2017-09-11 15:39:50.792177: W tensorflow/core/platform/cpu_feature_guard.cc:45] The TensorFlow librar
y wasn't compiled to use FMA instructions, but these are available on your machine and could speed up
 CPU computations.
>>> sess.run(a+b)
11
>>> sess.close()
>>> 
```

TensorFlow CPU installation for macOS X

In this section, we are going to install TensorFlow for macOS X using `virtualenv`. So, let's start off by installing the `pip` tool by issuing the following command:

```
sudo easy_install pip
```

Next, we need to install the virtual environment library:

```
sudo pip install --upgrade virtualenv
```

After installing the virtual environment library, we need to create a container or virtual environment which will host the installation of TensorFlow and any packages that you might want to install without affecting the underlying host system:

```
virtualenv --system-site-packages targetDirectory # for Python 2.7

virtualenv --system-site-packages -p python3 targetDirectory # for Python 3.n
```

This assumes that the `targetDirectory` is `~/tensorflow`.

Now that you have created the virtual environment, you can access it by issuing the following command:

```
source ~/tensorflow/bin/activate
```

Once you issue this command, you'll get access to the virtual machine that you have just created and you can install any packages that will be only installed in this environment and won't affect the underlying or host system that you're using.

In order to exit from the environment, you can issue the following command:

```
deactivate
```

Note that, for now, we do want to be inside the virtual environment, so turn it back on for now. Once you're done playing with TensorFlow, you should deactivate it:

```
source bin/activate
```

In order to install the CPU version of TensorFlow, you can issue the following commands, which will also install any dependent libraries that TensorFlow requires:

```
(tensorflow)$ pip install --upgrade tensorflow     # for Python 2.7

(tensorflow)$ pip3 install --upgrade tensorflow    # for Python 3.n
```

TensorFlow GPU/CPU installation for Windows

We will assume that you have Python 3 already installed on your system. To install TensorFlow, start a terminal as an administrator as follows. Open up the **Start** menu, search for **cmd,** and then right-click on it and click **Run as an administrator**:

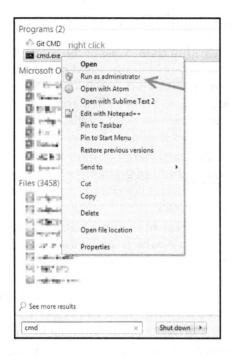

Once you have a command window opened, you can issue the following command to install TensorFlow in GPU mode:

 You need to have `pip` or `pip3` (depending on your Python version) installed before issuing the next command.

```
C:\> pip3 install --upgrade tensorflow-gpu
```

Issue the following command to install TensorFlow in CPU mode:

```
C:\> pip3 install --upgrade tensorflow
```

The TensorFlow environment

TensorFlow is another deep learning framework from Google and, as the name **TensorFlow** implies, it's derived from the operations which neural networks perform on multidimensional data arrays or tensors! It's literally a flow of tensors.

But first off, why are we going to use a deep learning framework in this book?

- **It scales machine learning code**: Most of the research into deep learning and machine learning can be applied/attributed because of these deep learning frameworks. They have allowed data scientists to iterate extremely quickly and have made deep learning and other ML algorithms much more accessible to practitioners. Big companies such as Google, Facebook, and so on are using such deep learning frameworks to scale to billions of users.
- **It computes gradients**: Deep learning frameworks can also compute gradients automatically. If you go through gradient calculation step by step, you will find out that gradient calculation is not trivial and it could be tricky to implement a bug-free version of it yourself.
- **It standardizes machine learning applications for sharing**: Also, pretrained models are available online, which can be used across different deep learning frameworks, and these pretrained models help people who have limited resources in terms of GPU so that they don't have to start from scratch every time. We can stand on the shoulders of giants and it take it from there.
- **There are lots of deep learning frameworks available** with different advantages, paradigms, levels of abstraction, programming languages, and so on.
- **Interface with GPUs for parallel processing**: Using GPUs for computations is a fascinating feature, because GPUs speed up your code a lot faster than the CPU because of number of cores and parallelization.

That's why Tensorflow is almost necessary in order to make progress in deep learning, because it can facilitate your projects.

So, briefly, what is TensorFlow?

- TensorFlow is a deep learning framework from Google which is open source for numerical computations using data flow graphs
- It was originally developed by the Google Brain Team to facilitate their machine learning research
- TensorFlow is an interface for expressing machine learning algorithms and an implementation for executing such algorithms

How does TensorFlow work and what is the underlying paradigm?

Computational graphs

The biggest idea of all of the big ideas about TensorFlow is that the numeric computations are expressed as a computation graph, as shown in the following figure. So, the backbone of any TensorFlow program is going to be a computational graph, where the following is true:

- Graph nodes are operations which have any number of inputs and outputs
- Graph edges between our nodes are going to be tensors that flow between these operations, and the best way of thinking about what tensors are in practice is as *n*-dimensional arrays

The advantage of using such flow graphs as the backbone of your deep learning framework is that it allows you to build complex models in terms of small and simple operations. Also, this is going to make the gradient calculations extremely simple when we address that in a later section:

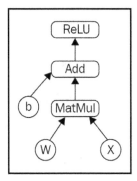

Another way of thinking about a TensorFlow graph is that each operation is a function that can be evaluated at that point.

TensorFlow data types, variables, and placeholders

The understanding of computational graphs will help us to think of complex models in terms of small subgraphs and operations.

Let's look at an example of a neural network with only one hidden layer and what its computation graph might look like in TensorFlow:

$$h = ReLU(Wx + b)$$

So, we have some hidden layer that we are trying to compute, as the ReLU activation of some parameter matrix W time some input x plus a bias term b. The ReLU function takes the max of your output and zero.

The following diagram shows what the graph might look like in TensorFlow:

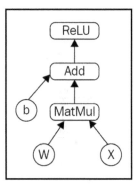

In this graph, we have variables for our b and W and we have something called a placeholder for x; we also have nodes for each of the operations in our graph. So, let's get into more detail about those node types.

Variables

Variables are going to be stateful nodes which output their current value. In this example, it's just b and W. What we mean by saying that variables are stateful is that they retain their current value over multiple executions and it's easy to restore saved values to variables:

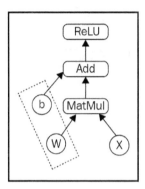

Also, variables have other useful features; for example, they can be saved to your disk during and after training, which facilities the use that we mentioned earlier that it allows people from different companies and groups to save, store, and send over their model parameters to other people. Also, the variables are the things that you want to tune to minimize the loss and we will see how to do that soon.

It's important to know that variables in the graph, such as b and W, are still operations because, by definition, all of your nodes in the graph are operations. So, when you evaluate these operations that are holding the values of b and W during runtime, you will get the value of those variables.

We can use the `Variable()` function of TensorFlow to define a variable and give it some initial value:

```
var = tf.Variable(tf.random_normal((0,1)),name='random_values')
```

This line of code will define a variable of 2 by 2 and initialize it from the standard normal distribution. You can also give a name to the variable.

Placeholders

The next type of nodes are placeholders. Placeholders are nodes whose values are fed at execution time:

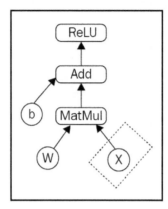

If you have inputs into your computational graph that depend on some external data, these are placeholders for values that we are going to add into our computation during training. So, for placeholders, we don't provide any initial values. We just assign a data type and shape of a tensor so the graph still knows what to compute even though it doesn't have any stored values yet.

We can use the placeholder function of TensorFlow to create a placeholder:

```
ph_var1 = tf.placeholder(tf.float32,shape=(2,3))
ph_var2 = tf.placeholder(tf.float32,shape=(3,2))
result = tf.matmul(ph_var1,ph_var2)
```

These lines of code define two placeholder variables of a certain shape and then define an operation (see the next section) that multiplies these two values together.

Mathematical operations

The third type of nodes are mathematical operations, and these are going to be our matrix multiplication (MatMul), addition (Add), and ReLU. All of these are nodes in your TensorFlow graph, and it's very similar to NumPy operations:

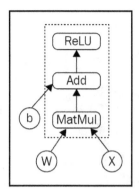

Let's see what this graph will look like in code.

We perform the following steps to produce the preceding graph:

1. Create weights W and b, including initialization. We can initialize the weight matrix W by sampling from uniform distribution $W \sim Uniform(-1,1)$ and initialize b to be 0.
2. Create input placeholder x, which is going to have a shape of $m * 784$ input matrix.
3. Build a flow graph.

Let's go ahead and follow those steps to build the flow graph:

```
# import TensorFlow package
import tensorflow as tf
# build a TensorFlow variable b taking in initial zeros of size 100
# ( a vector of 100 values)
b  = tf.Variable(tf.zeros((100,)))
# TensorFlow variable uniformly distributed values between -1 and 1
# of shape 784 by 100
W = tf.Variable(tf.random_uniform((784, 100),-1,1))
# TensorFlow placeholder for our input data that doesn't take in
# any initial values, it just takes a data type 32 bit floats as
# well as its shape
x = tf.placeholder(tf.float32, (100, 784))
# express h as Tensorflow ReLU of the TensorFlow matrix
```

```
#Multiplication of x and W and we add b
h = tf.nn.relu(tf.matmul(x,W) + b )
```

As you can see from the preceding code, we are not actually manipulating any data with this code snippet. We are only building symbols inside our graph and you can not print off *h* and see its value until we run this graph. So, this code snippet is just for building a backbone for our model. If you try to print the value of *W* or *b* in the preceding code, you should get the following output in Python:

```
ahmed@ahmed-Inspiron-7559:~$ python3
Python 3.6.0 |Anaconda 4.3.0 (64-bit)| (default, Dec 23 2016, 12:22:00)
[GCC 4.4.7 20120313 (Red Hat 4.4.7-1)] on linux
Type "help", "copyright", "credits" or "license" for more information.
>>> # import TensorFlow package
... import tensorflow as tf
>>> # build a TensorFlow variable b taking in initial zeros of size 100
... # ( a vector of 100 values)
... b  = tf.Variable(tf.zeros((100,)))
>>> # TensorFlow variable uniformly distributed values between -1 and 1
... # of shape 784 by 100
... W = tf.Variable(tf.random_uniform((784, 100),-1,1))
>>> # TensorFlow placeholder for our input data that doesn't take in
... # any initial values, it just takes a data type 32 bit floats as
... # well as its shape
... x = tf.placeholder(tf.float32, (100, 784))
>>> # express h as Tensorflow ReLU of the TensorFlow matrix
... #Multiplication of x and W and we add b
... h = tf.nn.relu(tf.matmul(x,W) + b )
>>> print(W)
<tf.Variable 'Variable_1:0' shape=(784, 100) dtype=float32_ref>
>>> print(b)
<tf.Variable 'Variable:0' shape=(100,) dtype=float32_ref>
>>> 
```

So far, we have defined our graph and now, we need to actually run it.

Getting output from TensorFlow

In the previous section, we knew how to build a computational graph, but we need to actually run it and get its value.

We can deploy/run the graph with something called a session, which is just a binding to a particular execution context such as a CPU or a GPU. So, we are going to take the graph that we build and deploy it to a CPU or a GPU context.

To run the graph, we need to define a session object called `sess`, and we are going to call the function `run` which takes two arguments:

```
sess.run(fetches, feeds)
```

Here:

- `fetches` are the list of the graph nodes that return the output of the nodes. These are the nodes we are interested in computing the value of.
- `feeds` are going to be a dictionary mapping from graph nodes to actual values that we want to run in our model. So, this is where we actually fill in the placeholders that we talked about earlier.

So, let's go ahead and run our graph:

```python
# importing the numpy package for generating random variables for
# our placeholder x
import numpy as np
# build a TensorFlow session object which takes a default execution
# environment which will be most likely a CPU
sess = tf.Session()
# calling the run function of the sess object to initialize all the
# variables.
sess.run(tf.global_variables_initializer())
# calling the run function on the node that we are interested in,
# the h, and we feed in our second argument which is a dictionary
# for our placeholder x with the values that we are interested in.
sess.run(h, {x: np.random.random((100,784))})
```

After running our graph through the `sess` object, we should get an output similar to the following:

```
>>> # importing the numpy package for generating random variables for
... # our placeholder x
... import numpy as np
>>> # build a TensorFlow session object which takes a default execution
... # environment which will be most likely a CPU
... sess = tf.Session()
>>> # calling the run function of the sess object to initialize all the
... # variables.
... sess.run(tf.global_variables_initializer())
>>> # calling the run function on the node that we are interested in,
... # the h, and we feed in our second argument which is a dictionary
... # for our placeholder x with the values that we are interested in.
... sess.run(h, {x: np.random.random((100,784))})
array([[ 4.95583916,  0.13156724,  0.         , ...,  0.          ,
         0.         ,  0.         ],
       [ 0.         ,  0.         ,  0.         , ...,  0.          ,
         0.         ,  0.         ],
       [ 0.         ,  0.         ,  0.         , ...,  0.          ,
         0.         ,  0.         ],
       ...,
       [ 2.81067681,  8.28696823,  0.         , ...,  0.          ,
         0.55022001,  0.         ],
       [ 0.         ,  6.67730427,  0.         , ...,  0.          ,
         4.28411198,  3.10559845],
       [ 1.92718267,  0.         ,  0.         , ...,  0.          ,
         0.         ,  0.         ]], dtype=float32)
>>> ▮
```

As you can see, in the second line of the above code snippet, we initialized our variables, and this is a concept in TensorFlow which is called **lazy evaluation**. It means that the evaluation of your graph only ever happens at runtime, and runtime in TensorFlow means the session. So, calling this function, `global_variables_initializer()`, will actually initialize anything called variable in your graph, such as *W* and *b* in our case.

We can also use the session variable in a with block to ensure that it will be closed after executing the graph:

```
ph_var1 = tf.placeholder(tf.float32,shape=(2,3))
ph_var2 = tf.placeholder(tf.float32,shape=(3,2))
result = tf.matmul(ph_var1,ph_var2)
with tf.Session() as sess:
print(sess.run([result],feed_dict={ph_var1:[[1.,3.,4.],[1.,3.,4.]],ph_var2:
[[1., 3.],[3.,1.],[.1,4.]]}))

Output:
[array([[10.4, 22. ],
        [10.4, 22. ]], dtype=float32)]
```

TensorBoard – visualizing learning

The computations you'll use TensorFlow for—such as training a massive deep neural network—can be complex and confusing, and its corresponding computational graph will be complex as well. To make it easier to understand, debug, and optimize TensorFlow programs, the TensorFlow team have included a suite of visualization tools called TensorBoard, which is a suite of web applications that can run through your browser. TensorBoard can be used to visualize your TensorFlow graph, plot quantitative metrics about the execution of your graph, and show additional data such as images that pass through it. When TensorBoard is fully configured, it looks like this:

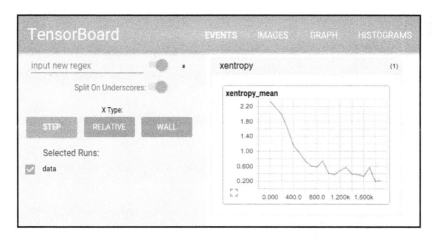

To understand how TensorBoard works, we are going to build a computational graph which will act as a classifier for the MNIST dataset, which is a dataset of handwritten images.

You don't have to understand all the bits and pieces of this model, but it will show you the general pipeline of a machine learning model implemented in TensorFlow.

So, let's start off by importing TensorFlow and loading the the required dataset using TensorFlow helper functions; these helper functions will check whether you're already downloaded the dataset, otherwise it will download it for you:

```
import tensorflow as tf

# Using TensorFlow helper function to get the MNIST dataset
from tensorflow.examples.tutorials.mnist import input_data
mnist_dataset = input_data.read_data_sets("/tmp/data/", one_hot=True)
```

```
Output:
Extracting /tmp/data/train-images-idx3-ubyte.gz
Extracting /tmp/data/train-labels-idx1-ubyte.gz
Extracting /tmp/data/t10k-images-idx3-ubyte.gz
Extracting /tmp/data/t10k-labels-idx1-ubyte.gz
```

Next, we need to define the hyperparameters (parameters that could be used to fine-tune the performance of your model) and inputs of our model:

```
# hyperparameters of the the model (you don't have to understand the
functionality of each parameter)
learning_rate = 0.01
num_training_epochs = 25
train_batch_size = 100
display_epoch = 1
logs_path = '/tmp/tensorflow_tensorboard/'

# Define the computational graph input which will be a vector of the image
pixels
# Images of MNIST has dimensions of 28 by 28 which will multiply to 784
input_values = tf.placeholder(tf.float32, [None, 784], name='input_values')

# Define the target of the model which will be a classification problem of
10 classes from 0 to 9
target_values = tf.placeholder(tf.float32, [None, 10],
name='target_values')

# Define some variables for the weights and biases of the model
weights = tf.Variable(tf.zeros([784, 10]), name='weights')
biases = tf.Variable(tf.zeros([10]), name='biases')
```

Now we need to build the model and define a cost function that we are going to optimize:

```
# Create the computational graph and encapsulating different operations to
different scopes
# which will make it easier for us to understand the visualizations of
TensorBoard
with tf.name_scope('Model'):
 # Defining the model
 predicted_values = tf.nn.softmax(tf.matmul(input_values, weights) +
biases)

with tf.name_scope('Loss'):
 # Minimizing the model error using cross entropy criteria
 model_cost = tf.reduce_mean(-
tf.reduce_sum(target_values*tf.log(predicted_values), reduction_indices=1))

with tf.name_scope('SGD'):
```

```
# using Gradient Descent as an optimization method for the model cost
above
model_optimizer =
tf.train.GradientDescentOptimizer(learning_rate).minimize(model_cost)

with tf.name_scope('Accuracy'):
 #Calculating the accuracy
 model_accuracy = tf.equal(tf.argmax(predicted_values, 1),
tf.argmax(target_values, 1))
 model_accuracy = tf.reduce_mean(tf.cast(model_accuracy, tf.float32))

# TensorFlow use the lazy evaluation strategy while defining the variables
# So actually till now none of the above variable got created or
initialized
init = tf.global_variables_initializer()
```

We will define the summary variable that will be used to monitor the changes that will happen on specific variables such as the loss and how it's getting better through out the training process:

```
# Create a summary to monitor the model cost tensor
tf.summary.scalar("model loss", model_cost)

# Create another summary to monitor the model accuracy tensor
tf.summary.scalar("model accuracy", model_accuracy)

# Merging the summaries to single operation
merged_summary_operation = tf.summary.merge_all()
```

Finally, we'll run the model by defining a session variable which will be used to execute the computation graph that we have built:

```
# kick off the training process
with tf.Session() as sess:

 # Intialize the variables
 sess.run(init)

 # operation to feed logs to TensorBoard
 summary_writer = tf.summary.FileWriter(logs_path,
graph=tf.get_default_graph())

 # Starting the training cycle by feeding the model by batch at a time
 for train_epoch in range(num_training_epochs):

 average_cost = 0.
 total_num_batch = int(mnist_dataset.train.num_examples/train_batch_size)
```

```
# iterate through all training batches
for i in range(total_num_batch):
batch_xs, batch_ys = mnist_dataset.train.next_batch(train_batch_size)

# Run the optimizer with gradient descent and cost to get the loss
# and the merged summary operations for the TensorBoard
_, c, summary = sess.run([model_optimizer, model_cost,
merged_summary_operation],
feed_dict={input_values: batch_xs, target_values: batch_ys})

# write statistics to the log et every iteration
summary_writer.add_summary(summary, train_epoch * total_num_batch + i)

# computing average loss
average_cost += c / total_num_batch

# Display logs per epoch step
if (train_epoch+1) % display_epoch == 0:
print("Epoch:", '%03d' % (train_epoch+1), "cost=",
"{:.9f}".format(average_cost))

print("Optimization Finished!")

# Testing the trained model on the test set and getting the accuracy
compared to the actual labels of the test set
print("Accuracy:", model_accuracy.eval({input_values:
mnist_dataset.test.images, target_values: mnist_dataset.test.labels}))

print("To view summaries in the Tensorboard, run the command line:\n" \
"--> tensorboard --logdir=/tmp/tensorflow_tensorboard " \
"\nThen open http://0.0.0.0:6006/ into your web browser")
```

The output of the training process should be similar to this:

```
Epoch: 001 cost= 1.183109128
Epoch: 002 cost= 0.665210275
Epoch: 003 cost= 0.552693334
Epoch: 004 cost= 0.498636444
Epoch: 005 cost= 0.465516675
Epoch: 006 cost= 0.442618381
Epoch: 007 cost= 0.425522513
Epoch: 008 cost= 0.412194222
Epoch: 009 cost= 0.401408134
Epoch: 010 cost= 0.392437336
Epoch: 011 cost= 0.384816745
Epoch: 012 cost= 0.378183398
Epoch: 013 cost= 0.372455584
Epoch: 014 cost= 0.367275238
```

```
Epoch: 015 cost= 0.362772711
Epoch: 016 cost= 0.358591895
Epoch: 017 cost= 0.354892231
Epoch: 018 cost= 0.351451424
Epoch: 019 cost= 0.348337946
Epoch: 020 cost= 0.345453095
Epoch: 021 cost= 0.342769080
Epoch: 022 cost= 0.340236065
Epoch: 023 cost= 0.337953151
Epoch: 024 cost= 0.335739001
Epoch: 025 cost= 0.333702818
Optimization Finished!
Accuracy: 0.9146
To view summaries in the Tensorboard, run the command line:
--> tensorboard --logdir=/tmp/tensorflow_tensorboard
Then open http://0.0.0.0:6006/ into your web browser
```

To see the summarized statistics in TensorBoard, we are going to follow the message at the end of our output by issuing the following command in the terminal:

```
tensorboard --logdir=/tmp/tensorflow_tensorboard
```

Then, open http://0.0.0.0:6006/ into your web browser.

When you open TensorBoard, you should get something similar to the following screenshot:

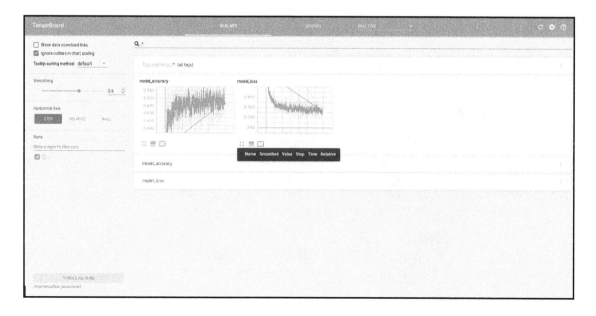

This shows the variables that we were monitoring, such as the model accuracy and how it's getting higher, and the model loss and how it's getting lower throughout the training process. So, you observe that we had a normal learning process here. But sometimes you will find out that the accuracy and model loss are changing randomly or you want to keep track of some variables and how they are changing throughout the session, and TensorBoard will be very useful to help you spot any randomness or mistakes.

Also, if switched to the **GRAPHS** tab in TensorBoard, you will get to see the computational graph that we built in the preceding code:

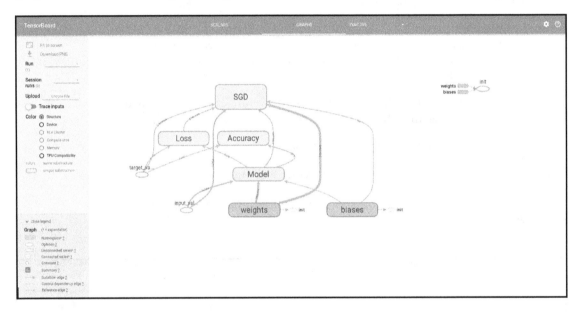

Summary

In this chapter, we covered the installation process for Ubuntu and Mac, gave an overview of the TensorFlow programming model, and explained the different types of simple nodes that could be used for building complex operations and how to get output from TensorFlow using a session object. Also, we covered TensorBoard and why it will helpful for debugging and analyzing complex deep learning applications.

Next, we will go through a basic explanation of neural networks and the intuition behind having multilayer neural networks. We will also cover some basic examples of TensorFlow and demonstrate how it could be used for regression and classification problems.

5
TensorFlow in Action - Some Basic Examples

,In this chapter, we will explain the main computational concept behind TensorFlow, which is the computational graph model, and demonstrate how to get you on track by implementing linear regression and logistic regression.

The following topics will be covered in this chapter:

- Capacity of a single neuron and activation functions
- Activation functions
- Feed-forward neural network
- The need for a multilayer network
- TensorFlow terminologies—recap
- Linear regression model—building and training
- Logistic regression model—building and training

We will start by explaining what a single neuron can actually do/model, and based on this, the need for a multilayer network will arise. Next up, we will do more elaboration of the main concepts and tools that are used/available within TensorFlow and how to use these tools to build up simple examples such as linear regression and logistic regression.

Capacity of a single neuron

A **neural network** is a computational model that is mainly inspired by the way the biological neural networks of the human brain process the incoming information. Neural networks made a huge breakthrough in machine learning research (deep learning, specifically) and industrial applications, such as breakthrough results in computer vision, speech recognition, and text processing. In this chapter, we will try to develop an understanding of a particular type of neural network called the **multi-layer Perceptron**.

Biological motivation and connections

The basic computational unit of our brains is called a **neuron**, and we have approximately 86 billion neurons in our nervous system, which are connected with approximately 10^{14} to 10^{15} synapses.

Figure 1 shows a biological neuron. *Figure 2* shows the corresponding mathematical model. In the drawing of the biological neuron, each neuron receives incoming signals from its dendrites and then produces output signals along its axon, where the axon gets split out and connects via synapses to other neurons.

In the corresponding mathematical computational model of a neuron, the signals that travel along the axons x_0 interact with a multiplication operation $w_0 x_0$ with the dendrites of the other neuron in the system based on the synaptic strength at that synapse, which is represented by w_0. The idea is that the synaptic weights/strength w gets learned by the network and they're the ones that control the influence of a specific neuron on another.

Also, in the basic computational model in *Figure 2*, the dendrites carry the signal to the main cell body where it sums them all. If the final result is above a certain threshold, the neuron can fire in the computational model.

Also, it is worth mentioning that we need to control the frequency of the output spikes along the axon, so we use something called an **activation function**. Practically, a common choice of activation function is the sigmoid function σ, since it takes a real-valued input (the signal strength after the sum) and squashes it to be between 0 and 1. We will see the details of these activation functions later in the following section:

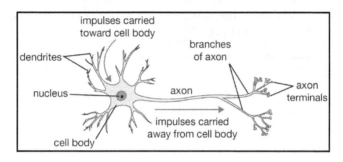

Figure 1: Computational unit of the brain (http://cs231n.github.io/assets/nn1/neuron.png)

There is the corresponding basic mathematical model for the biological one:

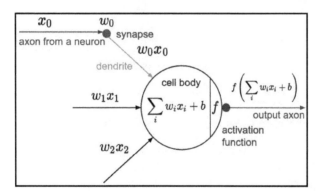

Figure 2: Mathematical modeling of the Brain's computational unit (http://cs231n.github.io/assets/nn1/neuron_model.jpeg)

The basic unit of computation in a neural network is the neuron, often called a **node** or **unit**. It receives input from some other nodes or from an external source and computes an output. Each input has an associated **weight** (**w**), which is assigned on the basis of its importance relative to other inputs. The node applies a function f (we've defined it later) to the weighted sum of its inputs.

So, the basic computational unit of neural networks in general is called **neuron/node/unit.**

This neuron receives its input from previous neurons or even an external source and then it does some processing on this input to produce a so-called activation. Each input to this neuron is associated with its own weight w, which represents the strength of this connection and hence the importance of this input.

So, the final output of this basic building block of the neural network is a summed version of the inputs weighted by their importance w, and then the neuron passes the summed output through an activation function.

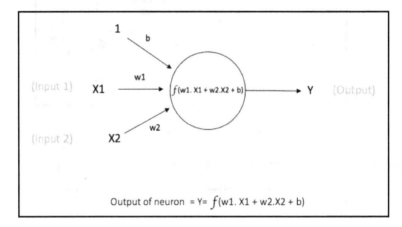

Figure 3: A single neuron

Activation functions

The output from the neuron is computed as shown in *Figure 3*, and passed through an activation function that introduces non-linearity to the output. This *f* is called an **activation function**. The main purposes of the activation functions are to:

- Introduce nonlinearity into the output of a neuron. This is important because most real-world data is nonlinear and we want neurons to learn these nonlinear representations.
- Squash the output to be in a specific range.

Every activation function (or nonlinearity) takes a single number and performs a certain fixed mathematical operation on it. There are several activation functions you may encounter in practice.

So, we are going to briefly cover the most common activation functions.

Sigmoid

Historically, the sigmoid activation function is widely used among researchers. This function accepts a real-valued input and squashes it to a range between 0 and 1, as shown in the following figure:

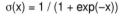

σ(x) = 1 / (1 + exp(−x))

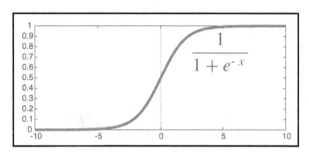

Figure 4: Sigmoid activation function

Tanh

Tanh is another activation function that tolerates some negative values. Tanh accepts a real-valued input and squashes them to [-1, 1]:

tanh(x) = 2σ(2x) − 1

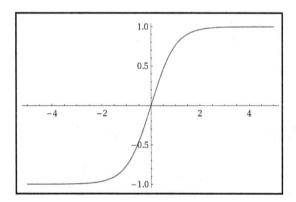

Figure 5: Tanh activation function

ReLU

Rectified linear unit (**ReLU**) does not tolerate negative values as it accepts a real-valued input and thresholds it at zero (replaces negative values with zero):

$$f(x) = max(0, x)$$

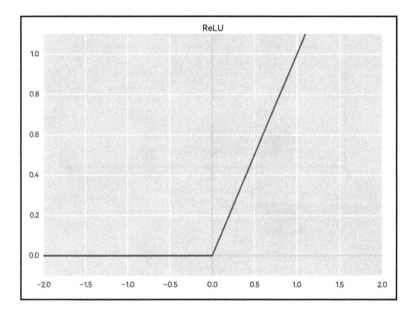

Figure 6: Relu activation function

Importance of bias: The main function of bias is to provide every node with a trainable constant value (in addition to the normal inputs that the node receives). See this link at `https://stackoverflow.com/questions/2480650/role-of-bias-in-neural-networks` to learn more about the role of bias in a neuron.

Feed-forward neural network

The feed-forward neural network was the first and simplest type of artificial neural network devised. It contains multiple neurons (nodes) arranged in layers. Nodes from adjacent layers have connections or edges between them. All these connections have weights associated with them.

An example of a feed-forward neural network is shown in *Figure 7*:

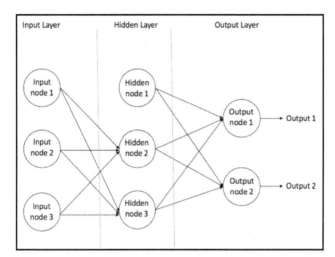

Figure 7: An example feed-forward neural network

In a feed-forward network, the information moves in only one direction—forward—from the input nodes, through the hidden nodes (if any), and to the output nodes. There are no cycles or loops in the network (this property of feed-forward networks is different from recurrent neural networks, in which the connections between nodes form a cycle).

The need for multilayer networks

A **multi-layer perceptron** (**MLP**) contains one or more hidden layers (apart from one input and one output layer). While a single layer perceptron can learn only linear functions, a MLP can also learn non-linear functions.

Figure 7 shows MLP with a single hidden layer. Note that all connections have weights associated with them, but only three weights ($w0$, $w1$, and $w2$) are shown in the figure.

Input Layer: The Input layer has three nodes. The bias node has a value of 1. The other two nodes take X1 and X2 as external inputs (which are numerical values depending upon the input dataset). As discussed before, no computation, is performed in the **Input Layer**, so the outputs from nodes in the **Input Layer** are **1**, **X1**, and **X2** respectively, which are fed into the **Hidden Layer**.

Hidden Layer: The **Hidden Layer** also has three nodes, with the bias node having an output of 1. The output of the other two nodes in the **Hidden Layer** depends on the outputs from the **Input Layer** (**1**, **X1**, and **X2**) as well as the weights associated with the connections (edges). Remember that *f* refers to the activation function. These outputs are then fed to the nodes in the **Output Layer**.

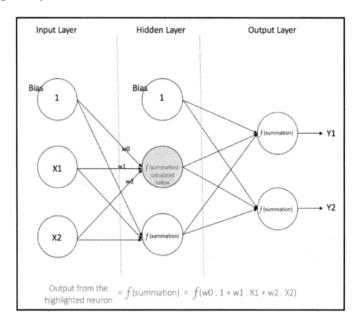

Figure 8: A multi-layer perceptron having one hidden layer

Output Layer: The **Output Layer** has two nodes; they take inputs from the **Hidden Layer** and perform similar computations as shown for the highlighted hidden node. The values calculated (**Y1** and **Y2**) as a result of these computations act as outputs of the multi-layer perceptron.

Given a set of features *X = (x1, x2, ...)* and a target *y*, a multi-layer perceptron can learn the relationship between the features and the target for either classification or regression.

Let's take an example to understand multi-layer perceptrons better. Suppose we have the following student marks dataset:

Table 1 – Sample student marks dataset

Hours studied	Mid term marks	Final term results
35	67	Pass
12	75	Fail
16	89	Pass
45	56	Pass
10	90	Fail

The two input columns show the number of hours the student has studied and the mid term marks obtained by the student. The **Final Result** column can have two values, **1** or **0**, indicating whether the student passed in the final term or not. For example, we can see that if the student studied 35 hours and had obtained 67 marks in the mid term, he/she ended up passing the final term.

Now, suppose we want to predict whether a student studying 25 hours and having 70 marks in the mid term will pass the final term:

Table 2 – Sample student with unknown final term result

Hours studied	Mid term marks	Final term result
26	70	?

This is a binary classification problem, where a MLP can learn from the given examples (training data) and make an informed prediction given a new data point. We will soon see how a MLP learns such relationships.

Training our MLP – the backpropagation algorithm

The process by which a multi-layer perceptron learns is called the **backpropagation** algorithm. I would recommend reading this Quora answer by Hemanth Kumar, `https://www.quora.com/How-do-you-explain-back-propagation-algorithm-to-a-beginner-in-neural-network/answer/Hemanth-Kumar-Mantri` (quoted later), which explains backpropagation clearly.

*"**Backward Propagation of Errors**, often abbreviated as BackProp is one of the several ways in which an artificial neural network (ANN) can be trained. It is a supervised training scheme, which means, it learns from labeled training data (there is a supervisor, to guide its learning).*

*To put in simple terms, BackProp is like "**learning from mistakes**". The supervisor corrects the ANN whenever it makes mistakes.*

An ANN consists of nodes in different layers; input layer, intermediate hidden layer(s) and the output layer. The connections between nodes of adjacent layers have "weights" associated with them. The goal of learning is to assign correct weights for these edges. Given an input vector, these weights determine what the output vector is.

In supervised learning, the training set is labeled. This means, for some given inputs, we know the desired/expected output (label).

BackProp Algorithm:

Initially all the edge weights are randomly assigned. For every input in the training dataset, the ANN is activated and its output is observed. This output is compared with the desired output that we already know, and the error is "propagated" back to the previous layer. This error is noted and the weights are "adjusted" accordingly. This process is repeated until the output error is below a predetermined threshold.

Once the above algorithm terminates, we have a "learned" ANN which, we consider is ready to work with "new" inputs. This ANN is said to have learned from several examples (labeled data) and from its mistakes (error propagation)."

—Hemanth Kumar.

Now that we have an idea of how backpropagation works, let's go back to our student marks dataset.

The MLP shown in *Figure 8* has two nodes in the input layer, which take the inputs hours studied and mid term marks. It also has a hidden layer with two nodes. The output layer has two nodes as well; the upper node outputs the probability of *pass* while the lower node outputs the probability of *fail*.

In classification applications, we widely use a softmax function (http://cs231n.github. io/linear-classify/#softmax) as the activation function in the output layer of the MLP to ensure that the outputs are probabilities and they add up to 1. The softmax function takes a vector of arbitrary real-valued scores and squashes it to a vector of values between 0 and 1 that sum up to 1. So, in this case:

$$Probability(Pass) + Probability(Fail) = 1$$

Step 1 – forward propagation

All weights in the network are randomly initialized. Let's consider a specific hidden layer node and call it *V*. Assume that the weights of the connections from the inputs to that node are **w1**, **w2**, and **w3** (as shown).

The network then takes the first training samples as input (we know that for inputs 35 and 67, the probability of passing is 1):

- Input to the network = [35, 67]
- Desired output from the network (target) = [1, 0]

Then, output *V* from the node in consideration, which can be calculated as follows (f is an activation function such as sigmoid):

$$V = f(1^*w1 + 35^*w2 + 67^*w3)$$

Similarly, outputs from the other node in the hidden layer are also calculated. The outputs of the two nodes in the hidden layer act as inputs to the two nodes in the output layer. This enables us to calculate output probabilities from the two nodes in the output layer.

Suppose the output probabilities from the two nodes in the output layer are 0.4 and 0.6, respectively (since the weights are randomly assigned, outputs will also be random). We can see that the calculated probabilities (0.4 and 0.6) are very far from the desired probabilities (1 and 0 respectively); hence the network is said to have an *incorrect output*.

Step 2 – backpropagation and weight updation

We calculate the total error at the output nodes and propagate these errors back through the network using backpropagation to calculate the gradients. Then, we use an optimization method such as gradient descent to adjust all weights in the network with an aim of reducing the error at the output layer.

Suppose that the new weights associated with the node in consideration are *w4*, *w5*, and *w6* (after backpropagation and adjusting weights).

If we now feed the same sample as an input to the network, the network should perform better than the initial run since the weights have now been optimized to minimize the error in prediction. The errors at the output nodes now reduce to [0.2, -0.2] as compared to [0.6, -0.4] earlier. This means that our network has learned to correctly classify our first training sample.

We repeat this process with all other training samples in our dataset. Then, our network is said to have learned those examples.

If we now want to predict whether a student studying 25 hours and having 70 marks in the mid term will pass the final term, we go through the forward propagation step and find the output probabilities for pass and fail.

I have avoided mathematical equations and explanation of concepts such as gradient descent here and have rather tried to develop an intuition for the algorithm. For a more mathematically involved discussion of the backpropagation algorithm, refer to this link: `http://home.agh.edu.pl/%7Evlsi/AI/backp_t_en/backprop.html`.

TensorFlow terminologies – recap

In this section, we will provide an overview of the TensorFlow library as well as the structure of a basic TensorFlow application. TensorFlow is an open source library for creating large-scale machine learning applications; it can model computations on a wide variety of hardware, ranging from android devices to heterogeneous multi-gpu systems.

TensorFlow uses a special structure in order to execute code on different devices such as CPUs and GPUs. Computations are defined as a graph and each graph is made up of operations, also known as **ops**, so whenever we work with TensorFlow, we define the series of operations in a graph.

To run these operations, we need to launch the graph into a session. The session translates the operations and passes them to a device for execution.

For example, the following image represents a graph in TensorFlow. *W*, *x*, and *b* are tensors over the edges of this graph. *MatMul* is an operation over the tensors *W* and *x*; after that, *Add* is called and we add the result of the previous operator with *b*. The resultant tensors of each operation cross the next one until the end, where it's possible to get the desired result.

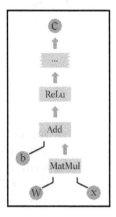

Figure 9: Sample TensorFlow computational graph

In order to use TensorFlow, we need to import the library; we'll give it the name `tf` so that we can access a module by writing `tf` dot and then the module's name:

```
import tensorflow as tf
```

To create our first graph, we will start by using source operations, which do not require any input. These source operations or source ops will pass their information to other operations, which will actually run computations.

Let's create two source operations that will output numbers. We will define them as `A` and `B`, which you can see in the following piece of code:

```
A = tf.constant([2])
```

```
B = tf.constant([3])
```

After that, we'll define a simple computational operation `tf.add()`, used to sum two elements. You can also use `C = A + B`, as shown in this code:

```
C = tf.add(A,B)
```

```
#C = A + B is also a way to define the sum of the terms
```

Since graphs need to be executed in the context of a session, we need to create a session object:

```
session = tf.Session()
```

To watch the graph, let's run the session to get the result from the previously defined C operation:

```
result = session.run(C)
print(result)

Output:
[5]
```

You're probably thinking that it was a lot of work just to add two numbers together, but it's extremely important that you understand the basic structure of TensorFlow. Once you do so, you can define any computations that you want; again, TensorFlow's structure allows it to handle computations on different devices (CPU or GPU), and even in clusters. If you want to learn more about this, you can run the method tf.device().

Also feel free to experiment with the structure of TensorFlow in order to get a better idea of how it works. If you want a list of all the mathematical operations that TensorFlow supports, you can check out the documentation.

By now, you should understand the structure of TensorFlow and how to create a basic applications.

Defining multidimensional arrays using TensorFlow

Now we will try to define such arrays using TensorFlow:

```
salar_var = tf.constant([4])
vector_var = tf.constant([5,4,2])
matrix_var = tf.constant([[1,2,3],[2,2,4],[3,5,5]])
tensor = tf.constant( [ [[1,2,3],[2,3,4],[3,4,5]] ,
[[4,5,6],[5,6,7],[6,7,8]] , [[7,8,9],[8,9,10],[9,10,11]] ] )
with tf.Session() as session:
    result = session.run(salar_var)
    print "Scalar (1 entry):\n %s \n" % result
    result = session.run(vector_var)
    print "Vector (3 entries) :\n %s \n" % result
    result = session.run(matrix_var)
    print "Matrix (3x3 entries):\n %s \n" % result
```

```
    result = session.run(tensor)
    print "Tensor (3x3x3 entries) :\n %s \n" % result
```

```
Output:
Scalar (1 entry):
 [2]

Vector (3 entries) :
 [5 6 2]

Matrix (3x3 entries):
 [[1 2 3]
 [2 3 4]
 [3 4 5]]

Tensor (3x3x3 entries) :
 [[[ 1  2  3]
  [ 2  3  4]
  [ 3  4  5]]

 [[ 4  5  6]
  [ 5  6  7]
  [ 6  7  8]]

 [[ 7  8  9]
  [ 8  9 10]
  [ 9 10 11]]]
```

Now that you understand these data structures, I encourage you to play with them using some previous functions to see how they will behave, according to their structure types:

```
Matrix_one = tf.constant([[1,2,3],[2,3,4],[3,4,5]])
Matrix_two = tf.constant([[2,2,2],[2,2,2],[2,2,2]])
first_operation = tf.add(Matrix_one, Matrix_two)
second_operation = Matrix_one + Matrix_two
with tf.Session() as session:
    result = session.run(first_operation)
    print "Defined using tensorflow function :"
    print(result)
    result = session.run(second_operation)
    print "Defined using normal expressions :"
    print(result)
```

```
Output:
Defined using tensorflow function :
[[3 4 5]
 [4 5 6]
 [5 6 7]]
```

```
Defined using normal expressions :
[[3 4 5]
 [4 5 6]
 [5 6 7]]
```

With the regular symbol definition and also the `tensorflow` function, we were able to get an element-wise multiplication, also known as **Hadamard product**. But what if we want the regular matrix product? We need to use another TensorFlow function called `tf.matmul()`:

```
Matrix_one = tf.constant([[2,3],[3,4]])
Matrix_two = tf.constant([[2,3],[3,4]])
first_operation = tf.matmul(Matrix_one, Matrix_two)
with tf.Session() as session:
    result = session.run(first_operation)
    print "Defined using tensorflow function :"
    print(result)
```

```
Output:
Defined using tensorflow function :
[[13 18]
 [18 25]]
```

We can also define this multiplication ourselves, but there is a function that already does that, so no need to reinvent the wheel!

Why tensors?

The tensor structure helps us by giving us the freedom to shape the dataset the way we want.

This is particularly helpful when dealing with images, due to the nature of how information in images are encoded.

Thinking about images, it's easy to understand that it has a height and width, so it would make sense to represent the information contained in it with a two-dimensional structure (a matrix)... until you remember that images have colors. To add information about the colors, we need another dimension, and that's when Tensors become particularly helpful.

Images are encoded into color channels; image data is represented in each color's intensity in a color channel at a given point, the most common one being RGB (which means red, blue, and green). The information contained in an image is the intensity of each channel color in the width and height of the image, just like this:

Figure 10: Different color channels for a specific image

So, the intensity of the red channel at each point with width and height can be represented in a matrix; the same goes for the blue and green channels. So, we end up having three matrices, and when these are combined, they form a tensor.

Variables

Now that we are more familiar with the structure of data, we will take a look at how TensorFlow handles variables.

To define variables, we use the command `tf.variable()`. To be able to use variables in a computation graph, it is necessary to initialize them before running the graph in a session. This is done by running `tf.global_variables_initializer()`.

To update the value of a variable, we simply run an assign operation that assigns a value to the variable:

```
state = tf.Variable(0)
```

Let's first create a simple counter, a variable that increases one unit at a time:

```
one = tf.constant(1)
new_value = tf.add(state, one)
update = tf.assign(state, new_value)
```

Variables must be initialized by running an initialization operation after having launched the graph. We first have to add the initialization operation to the graph:

```
init_op = tf.global_variables_initializer()
```

We then start a session to run the graph.

We first initialize the variables, then print the initial value of the state variable, and finally run the operation of updating the state variable and printing the result after each update:

```
with tf.Session() as session:
  session.run(init_op)
  print(session.run(state))
  for _ in range(3):
      session.run(update)
      print(session.run(state))

Output:
0
1
2
3
```

Placeholders

Now, we know how to manipulate variables inside TensorFlow, but what about feeding data outside of a TensorFlow model?

If you want to feed data to a TensorFlow model from outside a model, you will need to use placeholders.

So, what are these placeholders and what do they do? Placeholders can be seen as *holes* in your model, *holes* that you will pass the data to. You can create them using `tf.placeholder(datatype)`, where `datatype` specifies the type of data (integers, floating points, strings, and Booleans) along with its precision (8, 16, 32, and 64) bits.

The definition of each data type with the respective Python syntax is defined as:

Table 3 – Definition of different TensorFlow data types

Data type	Python type	Description
DT_FLOAT	tf.float32	32-bits floating point.
DT_DOUBLE	tf.float64	64-bits floating point
DT_INT8	tf.int8	8-bits signed integer.
DT_INT16	tf.int16	16-bits signed integer.

DT_INT32	tf.int32	32-bits signed integer.
DT_INT64	tf.int64	64-bits signed integer.
DT_UINT8	tf.uint8	8-bits unsigned integer.
DT_STRING	tf.string	Variable length byte arrays. Each element of a Tensor is a byte array.
DT_BOOL	tf.bool	Boolean.
DT_COMPLEX64	tf.complex64	Complex number made of two 32-bits floating points: real and imaginary parts.
DT_COMPLEX128	tf.complex128	Complex number made of two 64-bits floating points: real and imaginary parts.
DT_QINT8	tf.qint8	8-bits signed integer used in quantized ops.
DT_QINT32	tf.qint32	32-bits signed integer used in quantized ops.
DT_QUINT8	tf.quint8	8-bits unsigned integer used in quantized ops.

So let's create a placeholder:

```
a=tf.placeholder(tf.float32)
```

And define a simple multiplication operation:

```
b=a*2
```

Now, we need to define and run the session, but since we created a *hole* in the model to pass the data, when we initialize the session. We are obliged to pass an argument with the data; otherwise we get an error.

To pass the data to the model, we call the session with an extra argument, feed_dict, in which we should pass a dictionary with each placeholder name followed by its respective data, just like this:

```
with tf.Session() as sess:
    result = sess.run(b, feed_dict={a:3.5})
    print result

Output:
7.0
```

Since data in TensorFlow is passed in the form of multidimensional arrays, we can pass any kind of tensor through the placeholders to get the answer to the simple multiplication operation:

```
dictionary={a: [ [ [1,2,3],[4,5,6],[7,8,9],[10,11,12] ] , [
[13,14,15],[16,17,18],[19,20,21],[22,23,24] ] ] }
with tf.Session() as sess:
    result = sess.run(b,feed_dict=dictionary)
    print result
```

```
Output:
[[[  2.   4.   6.]
  [  8.  10.  12.]
  [ 14.  16.  18.]
  [ 20.  22.  24.]]

 [[ 26.  28.  30.]
  [ 32.  34.  36.]
  [ 38.  40.  42.]
  [ 44.  46.  48.]]]
```

Operations

Operations are nodes that represent mathematical operations over the tensors on a graph. These operations can be any kind of functions, like add and subtract tensors, or maybe an activation function.

`tf.matmul`, `tf.add`, and `tf.nn.sigmoid` are some of the operations in TensorFlow. These are like functions in Python, but operate directly over tensors and each one does a specific thing.

Other operations can be easily found at: `https://www.tensorflow.org/api_guides/python/math_ops`.

Let's play around with some of these operations:

```
a = tf.constant([5])
b = tf.constant([2])
c = tf.add(a,b)
d = tf.subtract(a,b)
with tf.Session() as session:
    result = session.run(c)
    print 'c =: %s' % result
    result = session.run(d)
    print 'd =: %s' % result
```

```
Output:
c =: [7]
d =: [3]
```

`tf.nn.sigmoid` is an activation function: it's a little more complicated, but this function helps learning models to evaluate what kind of information is good or not good.

Linear regression model – building and training

According to our explanation of linear regression in the `Chapter 2`, *Data Modeling in Action - The Titanic Example* we are going to rely on this definition to build a simple linear regression model.

Let's start off by importing the necessary packages for this implementation:

```
import numpy as np
import tensorflow as tf
import matplotlib.patches as mpatches
import matplotlib.pyplot as plt
plt.rcParams['figure.figsize'] = (10, 6)
```

Let's define an independent variable:

```
input_values = np.arange(0.0, 5.0, 0.1)
input_values

Output:
array([ 0. ,  0.1,  0.2,  0.3,  0.4,  0.5,  0.6,  0.7,  0.8,  0.9,  1. ,
        1.1,  1.2,  1.3,  1.4,  1.5,  1.6,  1.7,  1.8,  1.9,  2. ,  2.1,
        2.2,  2.3,  2.4,  2.5,  2.6,  2.7,  2.8,  2.9,  3. ,  3.1,  3.2,
        3.3,  3.4,  3.5,  3.6,  3.7,  3.8,  3.9,  4. ,  4.1,  4.2,  4.3,
        4.4,  4.5,  4.6,  4.7,  4.8,  4.9])

##You can adjust the slope and intercept to verify the changes in the graph
weight=1
bias=0
output = weight*input_values + bias
plt.plot(input_values,output)
plt.ylabel('Dependent Variable')
plt.xlabel('Indepdendent Variable')
plt.show()
Output:
```

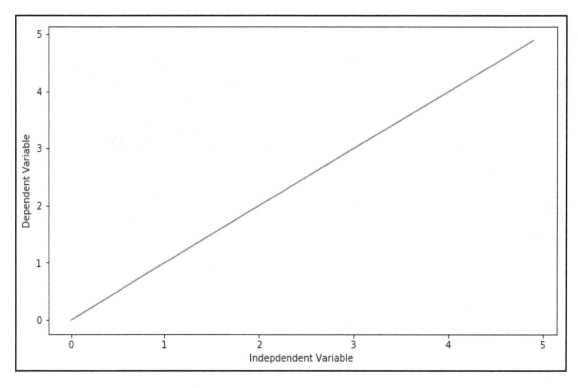

Figure 11: Visualization of the dependent variable versus the independent one

Now, let's see how this gets interpreted into a TensorFlow code.

Linear regression with TensorFlow

For the first part, we will generate random data points and define a linear relation; we'll use TensorFlow to adjust and get the right parameters:

```
input_values = np.random.rand(100).astype(np.float32)
```

The equation for the model used in this example is:

$$Y = 2X + 3$$

Nothing special about this equation, it is just a model that we use to generate our data points. In fact, you can change the parameters to whatever you want, as you will do later. We add some Gaussian noise to the points to make it a bit more interesting:

```
output_values = input_values * 2 + 3
output_values = np.vectorize(lambda y: y + np.random.normal(loc=0.0,
scale=0.1))(output_values)
```

Here is a sample of data:

```
list(zip(input_values,output_values))[5:10]

Output:
[(0.25240293, 3.474361759429548),
(0.946697, 4.980617375175061),
(0.37582186, 3.650345806087635),
(0.64025956, 4.271037640404975),
(0.62555283, 4.37001850440196)]
```

First, we initialize the variables *weight* and *bias* with any random guess, and then we define the linear function:

```
weight = tf.Variable(1.0)
bias = tf.Variable(0.2)
predicted_vals = weight * input_values + bias
```

In a typical linear regression model, we minimize the squared error of the equation that we want to adjust minus the target values (the data that we have), so we define the equation to be minimized as a loss.

To find loss's value, we use `tf.reduce_mean()`. This function finds the mean of a multidimensional tensor, and the result can have a different dimension:

```
model_loss = tf.reduce_mean(tf.square(predicted_vals - output_values))
```

Then, we define the optimizer method. Here, we will use a simple gradient descent with a learning rate of 0.5.

Now, we will define the training method of our graph, but what method will we use for minimize the loss? It's `tf.train.GradientDescentOptimizer`.

The `.minimize()` function will minimize the error function of our optimizer, resulting in a better model:

```
model_optimizer = tf.train.GradientDescentOptimizer(0.5)
train = model_optimizer.minimize(model_loss)
```

Don't forget to initialize the variables before executing a graph:

```
init = tf.global_variables_initializer()
sess = tf.Session()
sess.run(init)
```

Now, we are ready to start the optimization and run the graph:

```
train_data = []
for step in range(100):
    evals = sess.run([train,weight,bias])[1:]
    if step % 5 == 0:
        print(step, evals)
        train_data.append(evals)

Output:
(0, [2.5176678, 2.9857566])
(5, [2.4192538, 2.3015416])
(10, [2.5731843, 2.221911])
(15, [2.6890132, 2.1613526])
(20, [2.7763696, 2.1156814])
(25, [2.8422525, 2.0812368])
(30, [2.8919399, 2.0552595])
(35, [2.9294133, 2.0356679])
(40, [2.957675, 2.0208921])
(45, [2.9789894, 2.0097487])
(50, [2.9950645, 2.0013444])
(55, [3.0071881, 1.995006])
(60, [3.0163314, 1.9902257])
(65, [3.0232272, 1.9866205])
(70, [3.0284278, 1.9839015])
(75, [3.0323503, 1.9818509])
(80, [3.0353084, 1.9803041])
(85, [3.0375392, 1.9791379])
(90, [3.039222, 1.9782581])
(95, [3.0404909, 1.9775947])
```

Let's visualize the training process to fit the data points:

```
print('Plotting the data points with their corresponding fitted line...')
converter = plt.colors
cr, cg, cb = (1.0, 1.0, 0.0)

for f in train_data:

    cb += 1.0 / len(train_data)
    cg -= 1.0 / len(train_data)
```

```
if cb > 1.0: cb = 1.0

if cg < 0.0: cg = 0.0

[a, b] = f
f_y = np.vectorize(lambda x: a*x + b)(input_values)
line = plt.plot(input_values, f_y)
plt.setp(line, color=(cr,cg,cb))

plt.plot(input_values, output_values, 'ro')
green_line = mpatches.Patch(color='red', label='Data Points')
plt.legend(handles=[green_line])
plt.show()
```

Output:

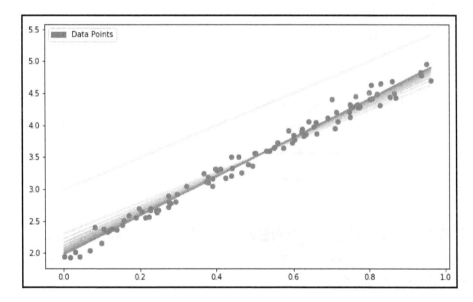

Figure 12: Visualization of the data points fitted by a regression line

Logistic regression model – building and training

Also based on our explanation of logistic regression in Chapter 2, *Data Modeling in Action - The Titanic Example*, we are going to implement the logistic regression algorithm in TensorFlow. So, briefly, logistic regression passes the input through the logistic/sigmoid but then treats the result as a probability:

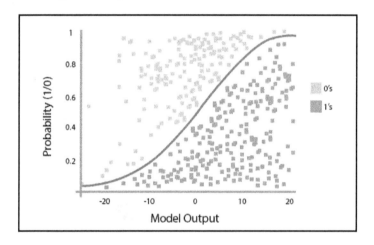

Figure 13: Discriminating between two linearly separable classes, 0's and 1's

Utilizing logistic regression in TensorFlow

For us to utilize logistic regression in TensorFlow, we first need to import whatever libraries we are going to use. To do so, you can run this code cell:

```
import tensorflow as tf

import pandas as pd

import numpy as np
import time
from sklearn.datasets import load_iris
from sklearn.cross_validation import train_test_split
import matplotlib.pyplot as plt
```

Next, we will load the dataset we are going to use. In this case, we are utilizing the iris dataset, which is inbuilt. So, there's no need to do any preprocessing and we can jump right into manipulating it. We separate the dataset into *x*'s and *y*'s, and then into training *x*'s and *y*'s and testing *x*'s and *y*'s, (pseudo) randomly:

```
iris_dataset = load_iris()
iris_input_values, iris_output_values = iris_dataset.data[:-1,:],
iris_dataset.target[:-1]
iris_output_values= pd.get_dummies(iris_output_values).values
train_input_values, test_input_values, train_target_values,
test_target_values = train_test_split(iris_input_values,
iris_output_values, test_size=0.33, random_state=42)
```

Now, we define *x* and *y*. These placeholders will hold our iris data (both the features and label matrices) and help pass them along to different parts of the algorithm. You can consider placeholders as empty shells into which we insert our data. We also need to give them shapes that correspond to the shape of our data. Later, we will insert data into these placeholders by feeding the placeholders the data via a `feed_dict` (feed dictionary):

Why use placeholders?

This feature of TensorFlow allows us to create an algorithm that accepts data and knows something about the shape of the data without knowing the amount of data going in. When we insert *batches* of data in training, we can easily adjust how many examples we train on in a single step without changing the entire algorithm:

```
# numFeatures is the number of features in our input data.
# In the iris dataset, this number is '4'.
num_explanatory_features = train_input_values.shape[1]

# numLabels is the number of classes our data points can be in.
# In the iris dataset, this number is '3'.
num_target_values = train_target_values.shape[1]


# Placeholders
# 'None' means TensorFlow shouldn't expect a fixed number in that dimension
input_values = tf.placeholder(tf.float32, [None, num_explanatory_features])
# Iris has 4 features, so X is a tensor to hold our data.
output_values = tf.placeholder(tf.float32, [None, num_target_values]) #
This will be our correct answers matrix for 3 classes.
```

Set model weights and bias

Much like linear regression, we need a shared variable weight matrix for logistic regression. We initialize both W and b as tensors full of zeros. Since we are going to learn W and b, their initial value doesn't matter too much. These variables are the objects that define the structure of our regression model, and we can save them after they've been trained so that we can reuse them later.

We define two TensorFlow variables as our parameters. These variables will hold the weights and biases of our logistic regression and they will be continually updated during training.

Notice that W has a shape of [4, 3] because we want to multiply the 4-dimensional input vectors by it to produce 3-dimensional vectors of evidence for the difference classes. b has a shape of [3], so we can add it to the output. Moreover, unlike our placeholders (which are essentially empty shells waiting to be fed data), TensorFlow variables need to be initialized with values, say, with zeros:

```
#Randomly sample from a normal distribution with standard deviation .01

weights =
tf.Variable(tf.random_normal([num_explanatory_features,num_target_values],
                             mean=0,
                             stddev=0.01,
                             name="weights"))

biases = tf.Variable(tf.random_normal([1,num_target_values],
                             mean=0,
                             stddev=0.01,
                             name="biases"))
```

Logistic regression model

We now define our operations in order to properly run the logistic regression. Logistic regression is typically thought of as a single equation:

$$\hat{y} = sigmoid(WX + b)$$

However, for the sake of clarity, we can have it broken into its three main components:

- A weight times features matrix multiplication operation
- A summation of the weighted features and a bias term
- Finally, the application of a sigmoid function

As such, you will find these components defined as three separate operations:

```
# Three-component breakdown of the Logistic Regression equation.
# Note that these feed into each other.
apply_weights = tf.matmul(input_values, weights, name="apply_weights")
add_bias = tf.add(apply_weights, biases, name="add_bias")
activation_output = tf.nn.sigmoid(add_bias, name="activation")
```

As we have seen previously, the function we are going to use is the logistic function, which is fed the input data after applying weights and bias. In TensorFlow, this function is implemented as the `nn.sigmoid` function. Effectively, it fits the weighted input with bias into a 0-100 percent curve, which is the probability function we want.

Training

The learning algorithm is how we search for the best weight vector (w). This search is an optimization problem looking for the hypothesis that optimizes an error/cost measure.

So, the cost or the loss function of the model is going to tell us our model is bad, and we need to minimize this function. There are different loss or cost criteria that you can follow. In this implementation, we are going to use **mean squared error** (**MSE**) as a loss function.

To accomplish the task of minimizing the loss function, we are going to use the gradient descent algorithm.

Cost function

Before defining our cost function, we need to define how long we are going to train and how we should define the learning rate:

```
#Number of training epochs
num_epochs = 700
# Defining our learning rate iterations (decay)
learning_rate = tf.train.exponential_decay(learning_rate=0.0008,
                                           global_step=1,
decay_steps=train_input_values.shape[0],
                                           decay_rate=0.95,
```

```
                                                    staircase=True)

# Defining our cost function - Squared Mean Error
model_cost = tf.nn.l2_loss(activation_output - output_values,
name="squared_error_cost")
# Defining our Gradient Descent
model_train =
tf.train.GradientDescentOptimizer(learning_rate).minimize(model_cost)
```

Now, it's time to execute our computational graph through the session variable.

So first off, we need to initialize our weights and biases with zeros or random values using `tf.initialize_all_variables()`. This initialization step will become a node in our computational graph, and when we put the graph into a session, the operation will run and create the variables:

```
# tensorflow session
sess = tf.Session()

# Initialize our variables.
init = tf.global_variables_initializer()
sess.run(init)

#We also want some additional operations to keep track of our model's
efficiency over time. We can do this like so:
# argmax(activation_output, 1) returns the label with the most probability
# argmax(output_values, 1) is the correct label
correct_predictions =
tf.equal(tf.argmax(activation_output,1),tf.argmax(output_values,1))

# If every false prediction is 0 and every true prediction is 1, the
average returns us the accuracy
model_accuracy = tf.reduce_mean(tf.cast(correct_predictions, "float"))

# Summary op for regression output
activation_summary = tf.summary.histogram("output", activation_output)

# Summary op for accuracy
accuracy_summary = tf.summary.scalar("accuracy", model_accuracy)

# Summary op for cost
cost_summary = tf.summary.scalar("cost", model_cost)

# Summary ops to check how variables weights and biases are updating after
each iteration to be visualized in TensorBoard
weight_summary = tf.summary.histogram("weights",
weights.eval(session=sess))
```

```
bias_summary = tf.summary.histogram("biases", biases.eval(session=sess))

merged = tf.summary.merge([activation_summary, accuracy_summary,
cost_summary, weight_summary, bias_summary])
writer = tf.summary.FileWriter("summary_logs", sess.graph)

#Now we can define and run the actual training loop, like this:
# Initialize reporting variables

inital_cost = 0
diff = 1
epoch_vals = []
accuracy_vals = []
costs = []

# Training epochs
for i in range(num_epochs):
    if i > 1 and diff < .0001:
        print("change in cost %g; convergence."%diff)
        break

    else:
        # Run training step
        step = sess.run(model_train, feed_dict={input_values:
train_input_values, output_values: train_target_values})

        # Report some stats evert 10 epochs
        if i % 10 == 0:
            # Add epoch to epoch_values
            epoch_vals.append(i)

            # Generate the accuracy stats of the model
            train_accuracy, new_cost = sess.run([model_accuracy,
model_cost], feed_dict={input_values: train_input_values, output_values:
train_target_values})

            # Add accuracy to live graphing variable
            accuracy_vals.append(train_accuracy)

            # Add cost to live graphing variable
            costs.append(new_cost)
>
            # Re-assign values for variables
            diff = abs(new_cost - inital_cost)
            cost = new_cost


            print("Training step %d, accuracy %g, cost %g, cost change
```

```
%g"%(i, train_accuracy, new_cost, diff))

Output:
Training step 0, accuracy 0.343434, cost 34.6022, cost change 34.6022
Training step 10, accuracy 0.434343, cost 30.3272, cost change 30.3272
Training step 20, accuracy 0.646465, cost 28.3478, cost change 28.3478
Training step 30, accuracy 0.646465, cost 26.6752, cost change 26.6752
Training step 40, accuracy 0.646465, cost 25.2844, cost change 25.2844
Training step 50, accuracy 0.646465, cost 24.1349, cost change 24.1349
Training step 60, accuracy 0.646465, cost 23.1835, cost change 23.1835
Training step 70, accuracy 0.646465, cost 22.3911, cost change 22.3911
Training step 80, accuracy 0.646465, cost 21.7254, cost change 21.7254
Training step 90, accuracy 0.646465, cost 21.1607, cost change 21.1607
Training step 100, accuracy 0.666667, cost 20.677, cost change 20.677
Training step 110, accuracy 0.666667, cost 20.2583, cost change 20.2583
Training step 120, accuracy 0.666667, cost 19.8927, cost change 19.8927
Training step 130, accuracy 0.666667, cost 19.5705, cost change 19.5705
Training step 140, accuracy 0.666667, cost 19.2842, cost change 19.2842
Training step 150, accuracy 0.666667, cost 19.0278, cost change 19.0278
Training step 160, accuracy 0.676768, cost 18.7966, cost change 18.7966
Training step 170, accuracy 0.69697, cost 18.5867, cost change 18.5867
Training step 180, accuracy 0.69697, cost 18.3951, cost change 18.3951
Training step 190, accuracy 0.717172, cost 18.2191, cost change 18.2191
Training step 200, accuracy 0.717172, cost 18.0567, cost change 18.0567
Training step 210, accuracy 0.737374, cost 17.906, cost change 17.906
Training step 220, accuracy 0.747475, cost 17.7657, cost change 17.7657
Training step 230, accuracy 0.747475, cost 17.6345, cost change 17.6345
Training step 240, accuracy 0.757576, cost 17.5113, cost change 17.5113
Training step 250, accuracy 0.787879, cost 17.3954, cost change 17.3954
Training step 260, accuracy 0.787879, cost 17.2858, cost change 17.2858
Training step 270, accuracy 0.787879, cost 17.182, cost change 17.182
Training step 280, accuracy 0.787879, cost 17.0834, cost change 17.0834
Training step 290, accuracy 0.787879, cost 16.9895, cost change 16.9895
Training step 300, accuracy 0.79798, cost 16.8999, cost change 16.8999
Training step 310, accuracy 0.79798, cost 16.8141, cost change 16.8141
Training step 320, accuracy 0.79798, cost 16.732, cost change 16.732
Training step 330, accuracy 0.79798, cost 16.6531, cost change 16.6531
Training step 340, accuracy 0.808081, cost 16.5772, cost change 16.5772
Training step 350, accuracy 0.818182, cost 16.5041, cost change 16.5041
Training step 360, accuracy 0.838384, cost 16.4336, cost change 16.4336
Training step 370, accuracy 0.838384, cost 16.3655, cost change 16.3655
Training step 380, accuracy 0.838384, cost 16.2997, cost change 16.2997
Training step 390, accuracy 0.838384, cost 16.2359, cost change 16.2359
Training step 400, accuracy 0.848485, cost 16.1741, cost change 16.1741
Training step 410, accuracy 0.848485, cost 16.1141, cost change 16.1141
Training step 420, accuracy 0.848485, cost 16.0558, cost change 16.0558
Training step 430, accuracy 0.858586, cost 15.9991, cost change 15.9991
Training step 440, accuracy 0.858586, cost 15.944, cost change 15.944
```

```
Training step 450, accuracy 0.858586, cost 15.8903, cost change 15.8903
Training step 460, accuracy 0.868687, cost 15.8379, cost change 15.8379
Training step 470, accuracy 0.878788, cost 15.7869, cost change 15.7869
Training step 480, accuracy 0.878788, cost 15.7371, cost change 15.7371
Training step 490, accuracy 0.878788, cost 15.6884, cost change 15.6884
Training step 500, accuracy 0.878788, cost 15.6409, cost change 15.6409
Training step 510, accuracy 0.878788, cost 15.5944, cost change 15.5944
Training step 520, accuracy 0.878788, cost 15.549, cost change 15.549
Training step 530, accuracy 0.888889, cost 15.5045, cost change 15.5045
Training step 540, accuracy 0.888889, cost 15.4609, cost change 15.4609
Training step 550, accuracy 0.89899, cost 15.4182, cost change 15.4182
Training step 560, accuracy 0.89899, cost 15.3764, cost change 15.3764
Training step 570, accuracy 0.89899, cost 15.3354, cost change 15.3354
Training step 580, accuracy 0.89899, cost 15.2952, cost change 15.2952
Training step 590, accuracy 0.909091, cost 15.2558, cost change 15.2558
Training step 600, accuracy 0.909091, cost 15.217, cost change 15.217
Training step 610, accuracy 0.909091, cost 15.179, cost change 15.179
Training step 620, accuracy 0.909091, cost 15.1417, cost change 15.1417
Training step 630, accuracy 0.909091, cost 15.105, cost change 15.105
Training step 640, accuracy 0.909091, cost 15.0689, cost change 15.0689
Training step 650, accuracy 0.909091, cost 15.0335, cost change 15.0335
Training step 660, accuracy 0.909091, cost 14.9987, cost change 14.9987
Training step 670, accuracy 0.909091, cost 14.9644, cost change 14.9644
Training step 680, accuracy 0.909091, cost 14.9307, cost change 14.9307
Training step 690, accuracy 0.909091, cost 14.8975, cost change 14.8975
```

Now, it's time to see how our trained model performs on the `iris` dataset, so let's test our trained model against the test set:

```
# test the model against the test set
print("final accuracy on test set: %s" %str(sess.run(model_accuracy,
feed_dict={input_values: test_input_values,
output_values: test_target_values}))

Output:
final accuracy on test set: 0.9
```

Getting 0.9 accuracy on the test set is really good and you can try to get better results by changing the number of epochs.

Summary

In this chapter, we went through a basic explanation of neural networks and and the need for multi-layer neural networks. We also covered the TensorFlow computational graph model with some basic examples, such as linear regression and logistic regression.

Next up, we will go through more advanced examples and demonstrate how TensorFlow can be used to build something like handwritten character recognition. We will also tackle the core idea of architecture engineering that has replaced feature engineering in traditional machine learning.

6

Deep Feed-forward Neural Networks - Implementing Digit Classification

A **feed-forward neural network** (**FNN**) is a special type of neural network wherein links/connections between neurons do not form a cycle. As such, it is different from other architectures in a neural network that we will get to study later on in this book (recurrent-type neural networks). The FNN is a widely used architecture and it was the first and simplest type of neural network.

In this chapter, we will go through the architecture of a typical ;FNN, and we will be using the TensorFlow library for this. After covering these concepts, we will give a practical example of digit classification. The question of this example is, *Given a set of images that contain handwritten digits, how can you classify these images into 10 different classes (0-9)?*

The following topics will be covered in this chapter:

- Hidden units and architecture design
- MNIST dataset analysis
- Digit classification - model building and training

Hidden units and architecture design

In the next section, we'll recap artificial neural networks; they can do a good job in classification tasks such as classifying handwritten digits.

Suppose we have the network shown in *Figure 1*:

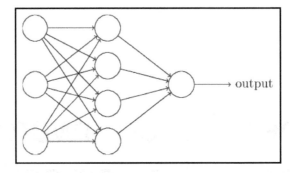

Figure 1: Simple FNN with one hidden layer

As mentioned earlier, the leftmost layer in this network is called the **input layer**, and the neurons within the layer are called **input neurons**. The rightmost or output layer contains the output neurons, or, as in this case, a single output neuron. The middle layer is called a **hidden layer**, since the neurons in this layer are neither inputs nor outputs. The term hidden perhaps sounds a little mysterious—the first time I heard the term, I thought it must have some deep philosophical or mathematical significance—but it really means *not an input and not an output*. It means nothing else. The preceding network has just a single hidden layer, but some networks have multiple hidden layers. For example, the following four-layer network has two hidden layers:

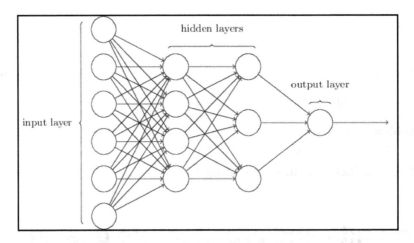

Figure 2: Artificial neural network with more hidden layers

The architecture in which the input, hidden, and output layers are organized is very straightforward. For example, let's go through a practical example to see whether a specific handwritten image has the digit 9 in it or not.

So first, we will feed the pixels of the input image to the input layer; for example, in the MNIST dataset, we have monochrome images. Each one of them is 28 by 28, so we need to have $28 \times 28 = 784$ neurons in the input layer to receive this input image.

In the output layer, we will need only 1 neuron, which produces a probability (or score) of whether this image has the digit 9 in it or not. For example, an output value of more than 0.5 means that this image has the digit 9, and if it's less than 0.5, then it means that the input image doesn't have the digit 9 in it.

So these types of networks, where the output from one layer is fed as an input to the next layer, are called FNNs. This kind of sequentiality in the layers means that there are no loops in it.

MNIST dataset analysis

In this section, we are going to get our hands dirty by implementing a classifier for handwritten images. This kind of implementation could be considered as the *Hello world!* of neural networks.

MNIST is a widely used dataset for benchmarking machine learning techniques. The dataset contains a set of handwritten digits like the ones shown here:

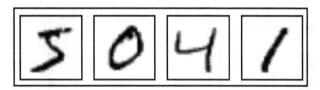

Figure 3: Sample digits from the MNIST dataset

So, the dataset includes handwritten images and their corresponding labels as well.

In this section, we are going to train a basic model on these images and the goal will be to tell which digit is handwritten in the input images.

Also, you'll find out that we will be able to accomplish this classification task using very few lines of code, but the idea behind this implementation is to understand the basic bits and pieces for building a neural network solution. Moreover, we are going to cover the main concepts of neural networking in this implementation.

The MNIST data

The MNIST data is hosted on Yann LeCun's website (http://yann.lecun.com/exdb/mnist/). Fortunately, TensorFlow provides some helper functions to download the dataset, so let's start off by downloading the dataset using the following two lines of code:

```
from tensorflow.examples.tutorials.mnist import input_data
mnist_dataset = input_data.read_data_sets("MNIST_data/", one_hot=True)
```

The MNIST data is split into three parts: 55,000 data points of training data (`mnist.train`), 10,000 points of test data (`mnist.test`), and 5,000 points of validation data (`mnist.validation`). This split is very important; it's essential in machine learning that we have separate data that we don't learn from so that we can make sure that what we've learned actually generalizes!

As mentioned earlier, every MNIST sample has two parts: an image of a handwritten digit and its corresponding label. Both the training set and test set contain images and their corresponding labels. For example, the training images are `mnist.train.images` and the training labels are `mnist.train.labels`.

Each image is 28 pixels by 28 pixels. We can interpret this as a big array of numbers:

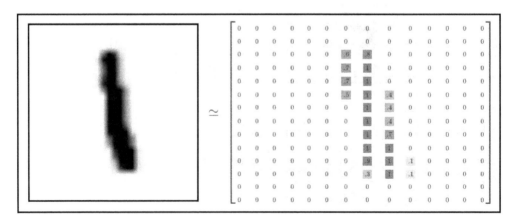

Figure 4: MNIST digit in matrix representation (intensity values)

In order to feed this matrix of pixel values to the input layer of the neural network, we need to flatten this matrix into a vector of 784 values. So, the final shape of the dataset will be a bunch of 784-dimensional vector space.

The result is that `mnist.train.images` is a tensor with a shape of `(55000, 784)`. The first dimension is an index of the list of images and the second dimension is the index for each pixel in each image. Each entry in the tensor is a pixel intensity between 0 and 1 for a particular pixel in a particular image:

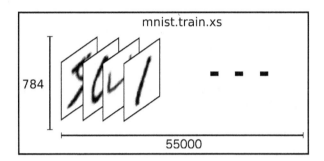

Figure 5: MNIST data analysis

As we mentioned previously, each image in the dataset has its corresponding label that ranges from 0 to 9.

For the purposes of this implementation, we're going to encode our labels as one-hot vectors. A one-hot vector is a vector of all zeros except the index of the digit that this vector represents. For example, 3 would be [0,0,0,1,0,0,0,0,0,0]. Consequently, `mnist.train.labels` is a `(55000, 10)` array of floats:

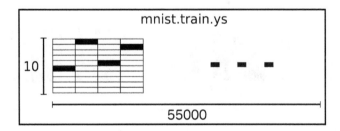

Figure 6: MNIST data analysis

Digit classification – model building and training

Now, let's go ahead and build our model. So, we have 10 classes in our dataset 0-9 and the goal is to classify any input image into one of these classes. Instead of giving a hard decision about the input image by saying only which class it could belong to, we are going to produce a vector of 10 possible values (because we have 10 classes). It'll represent the probabilities of each digit from 0-9 being the correct class for the input image.

For example, suppose we feed the model with a specific image. The model might be 70% sure that this image is 9, 10% sure that this image is 8, and so on. So, we are going to use the softmax regression here, which will produce values between 0 and 1.

A softmax regression has two steps: first we add up the evidence of our input being in certain classes, and then we convert that evidence into probabilities.

To tally up the evidence that a given image is in a particular class, we do a weighted sum of the pixel intensities. The weight is negative if that pixel having a high intensity is evidence against the image being in that class, and positive if it is evidence in favor.

Figure 7 shows the weights one model learned for each of these classes. Red represents negative weights, while blue represents positive weights:

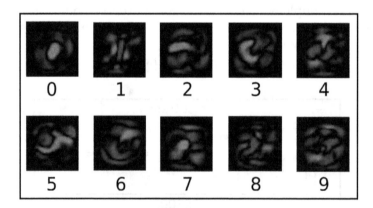

Figure 7: Weights one model learned for each of MNIST classes

We also add some extra evidence called a **bias**. Basically, we want to be able to say that some things are more likely independent of the input. The result is that the evidence for a class *i*, given an input, *x*, is:

$$evidence_i = \sum_j W_{i,j}\, x_j + b_i$$

Where:

- W_i is the weights
- b_i is the bias for class *i*
- *j* is an index for summing over the pixels in our input image *x*.

We then convert the evidence tallies into our predicted probabilities *y* using the softmax function:

$$y = softmax(evidence)$$

Here, softmax is serving as an activation or link function, shaping the output of our linear function into the form we want, in this case, a probability distribution over 10 cases (because we have 10 possible classes from 0-9). You can think of it as converting tallies of evidence into probabilities of our input being in each class. It's defined as:

$$softmax(evidence) = normalize(exp(evidence))$$

If you expand that equation, you get:

$$softmax(evidence)_i = \frac{exp(evidence_i))}{\sum_j exp(evidence_j)}$$

But it's often more helpful to think of softmax the first way: exponentiating its inputs and then normalizing them. Exponentiation means that one more unit of evidence increases the weight given to any hypothesis exponentially. And conversely, having one less unit of evidence means that a hypothesis gets a fraction of its earlier weight. No hypothesis ever has zero or negative weight. Softmax then normalizes these weights so that they add up to one, forming a valid probability distribution.

You can picture our softmax regression as looking something like the following, although with a lot more x's. For each output, we compute a weighted sum of the x's, add a bias, and then apply softmax:

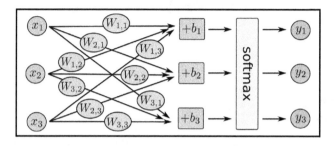

Figure 8: Visualization of softmax regression

If we write that out as equations, we get:

$$
\begin{bmatrix} y_1 \\ y_2 \\ y_3 \end{bmatrix} = \text{softmax} \begin{vmatrix} W_{1,1}x_1 + W_{1,2}x_2 + W_{1,3}x_3 + b_1 \\ W_{2,1}x_1 + W_{2,2}x_2 + W_{2,3}x_3 + b_2 \\ W_{3,1}x_1 + W_{3,2}x_2 + W_{3,3}x_3 + b_3 \end{vmatrix}
$$

Figure 9: Equation representation of the softmax regression

We can use vector notation for this procedure. This means that we'll be turning it into a matrix multiplication and vector addition. This is very helpful for computational efficiency and readability:

$$
\begin{bmatrix} y_1 \\ y_2 \\ y_3 \end{bmatrix} = \text{softmax} \begin{vmatrix} \begin{bmatrix} W_{1,1} & W_{1,2} & W_{1,3} \\ W_{2,1} & W_{2,2} & W_{2,3} \\ W_{3,1} & W_{3,2} & W_{3,3} \end{bmatrix} \cdot \begin{bmatrix} x_1 \\ x_2 \\ x_3 \end{bmatrix} + \begin{bmatrix} b_1 \\ b_2 \\ b_3 \end{bmatrix} \end{vmatrix}
$$

Figure 10: Vectorized representation of the softmax regression equation

More compactly, we can just write:

$$y = softmax(W_x + b)$$

Now, let's turn that into something that TensorFlow can use.

Data analysis

So, let's go ahead and start implementing our classifier. Let's start off by importing the required packages for this implementation:

```
import tensorflow as tf
import matplotlib.pyplot as plt
import numpy as np
import random as ran
```

Next up, we are going to define some helper functions to make us able to subset from the original dataset that we have downloaded:

```
#Define some helper functions
# to assign the size of training and test data we will take from MNIST
dataset
def train_size(size):
    print ('Total Training Images in Dataset = ' +
str(mnist_dataset.train.images.shape))
    print ('#############################################')
    input_values_train = mnist_dataset.train.images[:size,:]
    print ('input_values_train Samples Loaded = ' +
str(input_values_train.shape))
    target_values_train = mnist_dataset.train.labels[:size,:]
    print ('target_values_train Samples Loaded = ' +
str(target_values_train.shape))
    return input_values_train, target_values_train

def test_size(size):
    print ('Total Test Samples in MNIST Dataset = ' +
str(mnist_dataset.test.images.shape))
    print ('#############################################')
    input_values_test = mnist_dataset.test.images[:size,:]
    print ('input_values_test Samples Loaded = ' +
str(input_values_test.shape))
    target_values_test = mnist_dataset.test.labels[:size,:]
    print ('target_values_test Samples Loaded = ' +
str(target_values_test.shape))
    return input_values_test, target_values_test
```

Also, we're going to define two helper functions for displaying specific digits from the dataset or even display a flattened version of a subset of images:

```
#Define a couple of helper functions for digit images visualization
def visualize_digit(ind):
    print(target_values_train[ind])
    target = target_values_train[ind].argmax(axis=0)
    true_image = input_values_train[ind].reshape([28,28])
    plt.title('Sample: %d Label: %d' % (ind, target))
    plt.imshow(true_image, cmap=plt.get_cmap('gray_r'))
    plt.show()

def visualize_mult_imgs_flat(start, stop):
    imgs = input_values_train[start].reshape([1,784])
    for i in range(start+1,stop):
        imgs = np.concatenate((imgs,
input_values_train[i].reshape([1,784])))
    plt.imshow(imgs, cmap=plt.get_cmap('gray_r'))
    plt.show()
```

Now, let's get down to business and start playing around with the dataset. So we are going to define the training and testing examples that we would like to load from the original dataset.

Now, we'll get down to the business of building and training our model. First, we define variables with how many training and test examples we would like to load. For now, we will load all the data, but we will change this value later on to save resources:

```
input_values_train, target_values_train = train_size(55000)

Output:
Total Training Images in Dataset = (55000, 784)
###########################################
input_values_train Samples Loaded = (55000, 784)
target_values_train Samples Loaded = (55000, 10)
```

So now, we have a training set of 55,000 samples of handwritten digits, and each sample is 28 by 28 pixel images flattened to be a 784-dimensional vector. We also have their corresponding labels in a one-hot encoding format.

The `target_values_train` data are the associated labels for all the `input_values_train` samples. In the following example, the array represents a 7 in one-hot encoding format:

Label	0	1	2	3	4	5	6	7	8	9
Array	[0,	0,	0,	0,	0,	0,	0,	1,	0,	0]

Figure 11: One hot encoding for the digit 7

So let's visualize a random image from the dataset and see how it looks like, so we are going to use our preceding helper function to display a random digit from the dataset:

```
visualize_digit(ran.randint(0, input_values_train.shape[0]))
```

Output:

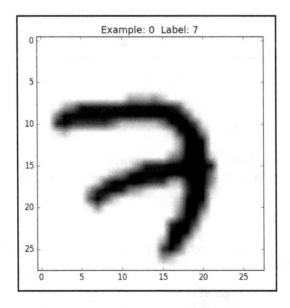

Figure 12: Output digit of the display_digit method

We can also visualize a bunch of flattened images using the helper function defined before. Each value in the flattened vector represents a pixel intensity, so visualizing the pixels will look like this:

```
visualize_mult_imgs_flat(0,400)
```

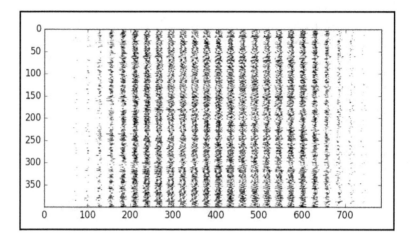

Figure 13: First 400 training examples

Building the model

So far, we haven't started to build our computational graph for this classifier. Let's start off by creating the session variable that will be responsible for executing the computational graph we are going to build:

```
sess = tf.Session()
```

Next up, we are going to define our model's placeholders, which will be used to feed data into the computational graph:

```
input_values = tf.placeholder(tf.float32, shape=[None, 784]
```

When we specify `None` in our placeholder's first dimension, it means the placeholder can be fed as many examples as we like. In this case, our placeholder can be fed any number of examples, where each example has a `784` value.

Now, we need to define another placeholder for feeding the image labels. Also we'll be using this placeholder later on to compare the model predictions with the actual labels of the images:

```
output_values = tf.placeholder(tf.float32, shape=[None, 10])
```

Next, we will define the `weights` and `biases`. These two variables will be the trainable parameters of our network and they will be the only two variables needed to make predictions on unseen data:

```
weights = tf.Variable(tf.zeros([784,10]))
biases = tf.Variable(tf.zeros([10]))
```

I like to think of these `weights` as 10 cheat sheets for each number. This is similar to how a teacher uses a cheat sheet to grade a multiple choice exam.

We will now define our softmax regression, which is our classifier function. This particular classifier is called **multinomial logistic regression**, and we make the prediction by multiplying the flattened version of the digit by the weight and then adding the bias:

```
softmax_layer = tf.nn.softmax(tf.matmul(input_values,weights) + biases)
```

First, let's ignore the softmax and look at what's inside the softmax function. `matmul` is the TensorFlow function for multiplying matrices. If you know matrix multiplication (https://en.wikipedia.org/wiki/Matrix_multiplication), you'll understand that this computes properly and that:

$$x \times W + b$$

Will result in a number of training examples fed (**m**) × number of classes (**n**) matrix:

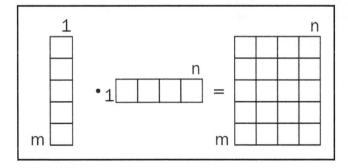

Figure 13: Simple matrix multiplication.

You can confirm it by evaluating `softmax_layer`:

```
print(softmax_layer)
Output:
Tensor("Softmax:0", shape=(?, 10), dtype=float32)
```

Now, let's experiment with the computational graph that we have defined previously with three samples from the training set and see how it works. To execute the computational graph, we need to use the session variable that we defined before. And we need to initialize the variables using `tf.global_variables_initializer()`.

Let's go ahead and only feed three samples to the computational graph:

```
input_values_train, target_values_train = train_size(3)
sess.run(tf.global_variables_initializer())
#If using TensorFlow prior to 0.12 use:
#sess.run(tf.initialize_all_variables())
print(sess.run(softmax_layer, feed_dict={input_values:
input_values_train}))

Output:

[[ 0.1   0.1   0.1   0.1   0.1   0.1   0.1   0.1   0.1   0.1]
 [ 0.1   0.1   0.1   0.1   0.1   0.1   0.1   0.1   0.1   0.1]
 [ 0.1   0.1   0.1   0.1   0.1   0.1   0.1   0.1   0.1   0.1]]
```

Here, you can see the model predictions for the three training samples that fed to it. At the moment, the model has learned nothing about our task because we haven't gone through the training process yet, so it just outputs 10% probability of each digit being the correct class for the input samples.

As we mentioned previously, softmax is an activation function that squashes the output to be between 0 and 1, and the TensorFlow implementation of softmax ensures that all the probabilities of a single input sample sums up to one.

Let's experiment a bit with the softmax function of TensorFlow:

```
sess.run(tf.nn.softmax(tf.zeros([4])))
sess.run(tf.nn.softmax(tf.constant([0.1, 0.005, 2])))

Output:
array([0.11634309, 0.10579926, 0.7778576 ], dtype=float32)
```

Next up, we need to define our loss function for this model, which will measure how good or bad our classifier is while trying to assign a class for the input images. The accuracy of our model is calculated by making a comparison between the actual values that we have in the dataset and the predictions that we got from the model.

The goal will be to reduce any misclassifications between the actual and predicted values.

Cross-entropy is defined as:

$$H_{y'}(y) = -\sum_i y_i' log(y_i)$$

Where:

- y is our predicted probability distribution
- y' is the true distribution (the one-hot vector with the digit labels)

In some rough sense, cross-entropy measures how inefficient our predictions are for describing the actual input.

We can implement the cross-entropy function:

```
model_cross_entropy = tf.reduce_mean(-tf.reduce_sum(output_values *
tf.log(softmax_layer), reduction_indices=[1]))
```

This function takes the log of all our predictions from `softmax_layer` (whose values range from 0 to 1) and multiplies them element-wise (https://en.wikipedia.org/wiki/Hadamard_product_%28matrices%29) by the example's true value, `output_values`. If the `log` function for each value is close to zero, it will make the value a large negative number (`-np.log(0.01)` = `4.6`), and if it is close to one, it will make the value a small negative number (`-np.log(0.99)` = `0.1`):

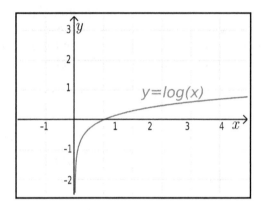

Figure 15: Visualization for Y = log (x)

We are essentially penalizing the classifier with a very large number if the prediction is confidently incorrect and a very small number if the prediction is confidently correct.

Here is a simple Python example of a softmax prediction that is very confident that the digit is a 3:

```
j = [0.03, 0.03, 0.01, 0.9, 0.01, 0.01, 0.0025,0.0025, 0.0025, 0.0025]
```

Let's create an array label of 3 as a ground truth to compare to our softmax function:

```
k = [0,0,0,1,0,0,0,0,0,0]
```

Can you guess what value our loss function gives us? Can you see how the log of `j` would penalize a wrong answer with a large negative number? Try this to understand:

```
-np.log(j)
-np.multiply(np.log(j),k)
```

This will return nine zeros and a value of 0.1053; when they all are summed up, we can consider this a good prediction. Notice what happens when we make the same prediction for what is actually a 2:

```
k = [0,0,1,0,0,0,0,0,0,0]
np.sum(-np.multiply(np.log(j),k))
```

Now, our `cross_entropy` function gives us 4.6051, which shows a heavily penalized, poorly made prediction. It was heavily penalized due to the fact the classifier was very confident that it was a 3 when it actually was a 2.

Next, we begin to train our classifier. In order to train it, we have to develop appropriate values for W and b that will give us the lowest possible loss.

The following is where we can now assign custom variables for training if we wish. Any value that is in all caps as follows is designed to be changed and messed with. In fact, I encourage it! First, use these values, and then notice what happens when you use too few training examples or too high or low of a learning rate:

```
input_values_train, target_values_train = train_size(5500)
input_values_test, target_values_test = test_size(10000)
learning_rate = 0.1
num_iterations = 2500
```

We can now initialize all variables so that they can be used by our TensorFlow graph:

```
init = tf.global_variables_initializer()
#If using TensorFlow prior to 0.12 use:
#init = tf.initialize_all_variables()
sess.run(init)
```

Next, we need to train the classifier using the gradient descent algorithm. So we first define our training method and some variables for measuring the model accuracy. The variable `train` will perform the gradient descent optimizer with a chosen learning rate in order to minimize the model loss function `model_cross_entropy`:

```
train =
tf.train.GradientDescentOptimizer(learning_rate).minimize(model_cross_entro
py)
model_correct_prediction = tf.equal(tf.argmax(softmax_layer,1),
tf.argmax(output_values,1))
model_accuracy = tf.reduce_mean(tf.cast(model_correct_prediction,
tf.float32))
```

Model training

Now, we'll define a loop that iterates `num_iterations` times. And for each loop, it runs training, feeding in values from `input_values_train` and `target_values_train` using `feed_dict`.

In order to calculate accuracy, it will test the model against the unseen data in `input_values_test`:

```
for i in range(num_iterations+1):
    sess.run(train, feed_dict={input_values: input_values_train,
output_values: target_values_train})
    if i%100 == 0:
        print('Training Step:' + str(i) + ' Accuracy = ' +
str(sess.run(model_accuracy, feed_dict={input_values: input_values_test,
output_values: target_values_test})) + ' Loss = ' +
str(sess.run(model_cross_entropy, {input_values: input_values_train,
output_values: target_values_train})))

Output:
Training Step:0 Accuracy = 0.5988 Loss = 2.1881988
Training Step:100 Accuracy = 0.8647 Loss = 0.58029664
Training Step:200 Accuracy = 0.879 Loss = 0.45982164
Training Step:300 Accuracy = 0.8866 Loss = 0.40857208
Training Step:400 Accuracy = 0.8904 Loss = 0.37808096
Training Step:500 Accuracy = 0.8943 Loss = 0.35697535
Training Step:600 Accuracy = 0.8974 Loss = 0.34104997
Training Step:700 Accuracy = 0.8984 Loss = 0.32834956
Training Step:800 Accuracy = 0.9 Loss = 0.31782663
Training Step:900 Accuracy = 0.9005 Loss = 0.30886236
Training Step:1000 Accuracy = 0.9009 Loss = 0.3010645
Training Step:1100 Accuracy = 0.9023 Loss = 0.29417014
Training Step:1200 Accuracy = 0.9029 Loss = 0.28799513
Training Step:1300 Accuracy = 0.9033 Loss = 0.28240603
Training Step:1400 Accuracy = 0.9039 Loss = 0.27730304
Training Step:1500 Accuracy = 0.9048 Loss = 0.27260992
Training Step:1600 Accuracy = 0.9057 Loss = 0.26826677
Training Step:1700 Accuracy = 0.9062 Loss = 0.2642261
Training Step:1800 Accuracy = 0.9061 Loss = 0.26044932
Training Step:1900 Accuracy = 0.9063 Loss = 0.25690478
Training Step:2000 Accuracy = 0.9066 Loss = 0.2535662
Training Step:2100 Accuracy = 0.9072 Loss = 0.25041154
Training Step:2200 Accuracy = 0.9073 Loss = 0.24742197
Training Step:2300 Accuracy = 0.9071 Loss = 0.24458146
Training Step:2400 Accuracy = 0.9066 Loss = 0.24187621
Training Step:2500 Accuracy = 0.9067 Loss = 0.23929419
```

Notice how the loss was still decreasing near the end but our accuracy slightly went down! This shows that we can still minimize our loss and hence maximize the accuracy over our training data, but this may not help us predict the testing data used for measuring accuracy. This is also known as **overfitting** (not generalizing). With the default settings, we got an accuracy of about 91%. If I wanted to cheat to get 94% accuracy, I could've set the test examples to 100. This shows how not having enough test examples can give you a biased sense of accuracy.

Keep in mind that this is a very inaccurate way to calculate our classifier's performance. However, we did this on purpose for the sake of learning and experimentation. Ideally, when training with large datasets, you train using small batches of training data at a time, not all at once.

This is the interesting part. Now that we have calculated our weight cheat sheet, we can create a graph with the following code:

```
for i in range(10):
    plt.subplot(2, 5, i+1)
    weight = sess.run(weights)[:,i]
    plt.title(i)
    plt.imshow(weight.reshape([28,28]), cmap=plt.get_cmap('seismic'))
    frame = plt.gca()
    frame.axes.get_xaxis().set_visible(False)
    frame.axes.get_yaxis().set_visible(False)
```

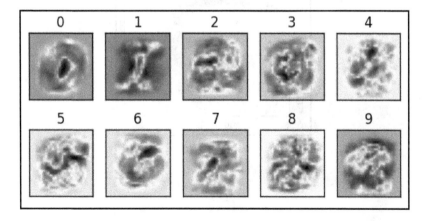

Figure 15: Visualization of our weights from 0-9

The preceding figure shows the model weights from 0-9, which is the most important side of our classifier. All this amount of work of machine learning is done to figure out what the optimal weights are. Once they are calculated based on an optimization criteria, you have the **cheat sheet** and can easily find your answers using the learned weights.

The learned model makes its prediction by comparing how similar or different the input digit sample is to the red and blue weights. The darker the red, the better the hit; white means neutral and blue means misses.

Now, let's use the cheat sheet and see how our model performs on it:

```
input_values_train, target_values_train = train_size(1)
visualize_digit(0)

Output:
Total Training Images in Dataset = (55000, 784)
############################################
input_values_train Samples Loaded = (1, 784)
target_values_train Samples Loaded = (1, 10)
[0. 0. 0. 0. 0. 0. 0. 1. 0. 0.]
```

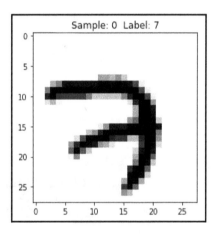

Let's look at our softmax predictor:

```
answer = sess.run(softmax_layer, feed_dict={input_values:
input_values_train})
print(answer)
```

The preceding code will give us a 10-dimensional vector, with each column containing one probability:

```
[[2.1248012e-05 1.1646927e-05 8.9631692e-02 1.9201526e-02 8.2086492e-04
  1.2516821e-05 3.8538201e-05 8.5374612e-01 6.9188857e-03 2.9596921e-02]]
```

We can use the `argmax` function to find out the most probable digit to be the correct classification for our input image:

```
answer.argmax()
```

```
Output:
7
```

Now, we get a correct classification from our network.

Let's use our knowledge to define a helper function that can select a random image from the dataset and test the model against it:

```
def display_result(ind):
    # Loading a training sample
    input_values_train = mnist_dataset.train.images[ind,:].reshape(1,784)
    target_values_train = mnist_dataset.train.labels[ind,:]
    # getting the label as an integer instead of one-hot encoded vector
    label = target_values_train.argmax()
    # Getting the prediction as an integer
    prediction = sess.run(softmax_layer, feed_dict={input_values:
input_values_train}).argmax()
    plt.title('Prediction: %d Label: %d' % (prediction, label))
    plt.imshow(input_values_train.reshape([28,28]),
cmap=plt.get_cmap('gray_r'))
    plt.show()
```

And now try it out:

```
display_result(ran.randint(0, 55000))
```

Output:

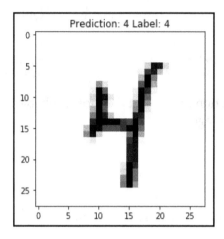

We've got a correct classification again!

Summary

In this chapter, we went through a basic implementation of a FNN for our digit classification task. We also did a recap of the terminologies used in the neural network context.

Next up, we will build a sophisticated version of the digit classification model using some modern best practices and some tips and tricks to enhance the model's performance.

7
Introduction to Convolutional Neural Networks

In data science, a **convolutional neural network** (**CNN**) is specific kind of deep learning architecture that uses the convolution operation to extract relevant explanatory features for the input image. CNN layers are connected as a feed-forward neural network while using this convolution operation to mimic how the human brain functions while trying to recognize objects. Individual cortical neurons respond to stimuli in a restricted region of space known as the receptive field. In particular, biomedical imaging problems could be challenging sometimes, but in this chapter, we'll see how to use CNN in order to discover patterns in this image.

The following topics will be covered in this chapter:

- The convolution operation
- Motivation
- Different layers of CNNs
- CNN basic example: MNIST digit classification

The convolution operation

CNNs are widely used in the area of computer vision and they outperform most of the traditional computer vision techniques that we have been using. CNNs combine the famous convolution operation and neural networks, hence the name convolutional neural network. So, before diving into the neural network aspect of CNNs, we are going to introduce the convolution operation and see how it works.

The main purpose of the convolution operation is to extract information or features from an image. Any image could be considered as a matrix of values and a specific group of values in this matrix will form a feature. The purpose of the convolution operation is to scan this matrix and try to extract relevant or explanatory features for that image. For example, consider a 5 by 5 image whose corresponding intensity or pixel values are shown as zeros and ones:

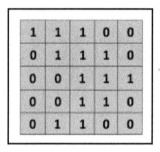

Figure 9.1: Matrix of pixel values

And consider the following 3 x 3 matrix:

Figure 9.2: Matrix of pixel values

We can convolve the 5 x 5 image using a 3 x 3 one as follows:

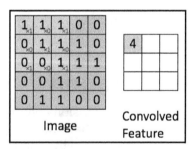

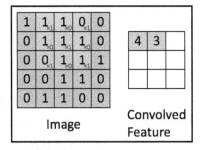

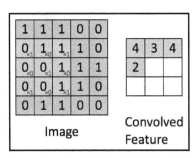

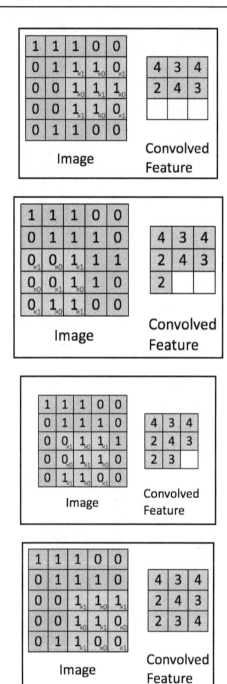

Figure 9.3: The convolution operation. The output matrix is called a convolved feature or feature map

The preceding figure could be summarized as follows. In order to convolve the original 5 by 5 image using the 3 x 3 convolution kernel, we need to do the following:

- Scan the original green image using the orange matrix and each time move by only 1 pixel (stride)
- For every position of the orange image, we do element-wise multiplication between the orange matrix and the corresponding pixel values in the green matrix
- Add the results of these element-wise multiplication operations together to get a single integer which will form a single value in the output pink matrix

 As you can see from the preceding figure, the orange 3 by 3 matrix only operates on one part of the original green image at a time in each move (stride), or it only sees a part at a time.

So, let's put the previous explanation in the context of CNN terminology:

- The orange 3 x 3 matrix is called a **kernel**, **feature detector**, or **filter**
- The output pink matrix that contain the results of the element-wise multiplications is called the **feature map**

Because of the fact that we are getting the feature map based on the element-wise multiplication between the kernel and the corresponding pixels in the original input image, changing the values of the kernel or the filter will give different feature maps each time.

So, we might think that we need to figure out the values of the feature detectors ourselves during the training of the convolution neural networks, but this is not the case here. CNNs figure out these numbers during the learning process. So, if we have more filters, it means that we can extract more features from the image.

Before jumping to the next section, let's introduce some terminology that is usually used in the context of CNNs:

- **Stride**: We mentioned this term briefly earlier. In general, stride is the number of pixels by which we move our feature detector or filter over the pixels of the input matrix. For example, stride 1 means moving the filter one pixel at a time while convolving the input image and stride 2 means moving the filter two pixels at a time while convolving the input image. The more stride we have, the smaller the generated feature maps are.
- **Zero-padding**: If we wanted to include the border pixels of the input image, then part of our filter will be outside the input image. Zero-padding solves this problem by padding the input matrix with zeros around the borders.

Motivation

Traditional computer vision techniques were used to perform most computer vision tasks, such as object detection and segmentation. The performance of these traditional computer vision techniques was good but it was never close to being usable in real time, for example by autonomous cars. In 2012, Alex Krizhevsky introduced CNNs, which made a breakthrough on the ImageNet competition by enhancing the object classification error from 26% to 15%. CNNs have been widely used since then and different variations have been discovered. It has even outperformed the human classification error over the ImageNet competition, as shown in the following diagram:

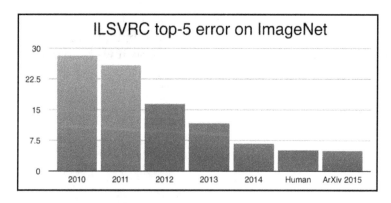

Figure 9.4: Classification error over time with human level error marked in red

Applications of CNNs

Since the breakthrough the CNNs achieved in different domains of computer vision and even natural language processing, most companies have integrated this deep learning solution into their computer vision echo system. For example, Google uses this architecture for its image search engine, and Facebook uses it for doing automatic tagging and more:

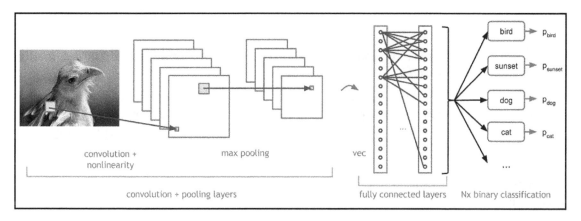

Figure 9.5: A typical CNN general architecture for object recognition

CNNs achieved this breakthrough because of their architecture, which intuitively uses the convolution operation to extract features from the images. Later on, you will see that it's very similar to the way the human brain works.

Different layers of CNNs

A typical CNN architecture consists of multiple layers that do different tasks, as shown in the preceding diagram. In this section, we are going to go through them in detail and will see the benefits of having all of them connected in a special way to make such a breakthrough in computer vision.

Input layer

This is the first layer in any CNN architecture. All the subsequent convolution and pooling layers expect the input to be in a specific format. The input variables will tensors, that has the following shape:

```
[batch_size, image_width, image_height, channels]
```

Here:

- `batch_size` is a random sample from the original training set that's used during applying stochastic gradient descent.
- `image_width` is the width of the input images to the network.
- `image_height` is the height of the input images to the network.
- `channels` are the number of color channels of the input images. This number could be 3 for RGB images or 1 for binary images.

For example, consider our famous MNIST dataset. Let's say we are going to perform digit classification using CNNs using this dataset.

If the dataset is composed of monochrome 28 x 28 pixel images like the MNIST dataset, then the desired shape for our input layer is as follows:

```
[batch_size, 28, 28, 1].
```

To change the shape of our input features, we can do the following reshaping operation:

```
input_layer = tf.reshape(features["x"], [-1, 28, 28, 1])
```

As you can see, we have specified the batch size to be -1, which means that this number should be determined dynamically based on the input values in the features. By doing this, we will be able to fine-tune our CNN model by controlling the batch size.

As an example for the reshape operation, suppose that we divided our input samples into batches of five and our feature `["x"]` array will hold 3,920 `values()` of the input images, where each value of this array corresponds to a pixel in an image. For this case, the input layer will have the following shape:

```
[5, 28, 28, 1]
```

Convolution step

As mentioned earlier, the convolution step got its name from the convolution operation. The main purpose of having these convolution steps is to extract features from the input images and then feed them to a linear classifier.

In natural images, features could be anywhere in the image. For example, edges could be in the middle or at the corner of the images, so the whole idea of stacking a bunch of convolution steps is to be able to detect these features anywhere in the image.

It's very easy to define a convolution step in TensorFlow. For example, if we wanted to apply 20 filters each of size 5 by 5 to the input layer with a ReLU activation function, then we can use the following line of code to do that:

```
conv_layer1 = tf.layers.conv2d(
  inputs=input_layer,
  filters=20,
  kernel_size=[5, 5],
  padding="same",
  activation=tf.nn.relu)
```

The first parameter for this `conv2d` function is the input layer that we have defined in the preceding code, which has the appropriate shape, and the second argument is the filters argument which specifies the number of filters to be applied to the image where the higher the number of filters, the more features are extracted from the input image. The third parameter is the `kernel_size`, which represents the size of the filter or the feature detector. The padding parameters specifies where we use `"same"` here to introduce zero-padding to the corner pixels of the input image. The last argument specifies the activation function that should be used for the output of the convolution operation.

So, in our MNIST example, the input tensor will have the following shape:

```
[batch_size, 28, 28, 1]
```

And the output tensor for this convolution step will have the following shape:

```
[batch_size, 28, 28, 20]
```

The output tensor has the same dimensions as the input images, but now we have 20 channels that represent applying the 20 filters to the input image.

Introducing non-linearity

In the convolution step, we talked about feeding the output of the convolution step to a ReLU activation function to introduce non-linearity:

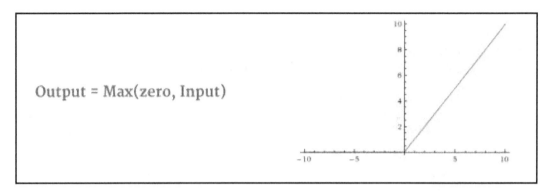

Figure 9.6: ReLU activation function

The ReLU activation function replaces all the negative pixel values with zeros and the whole purpose of feeding the output of the convolution step to this activation function is to introduce non-linearity in the output image because this will be useful for the training process as the data that we are using is usually non-linear. To clearly understand the benefit of ReLU activation function, have a look at the following figure, which shows the row output of the convolution step and the rectified version of it:

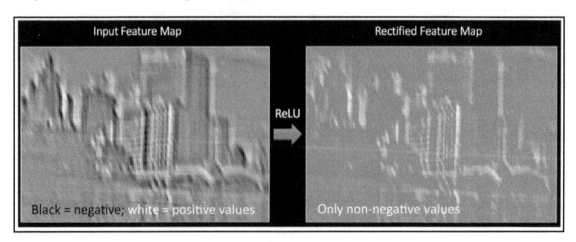

Figure 9.7: The result of applying ReLU to the input feature map

The pooling step

One of the important steps for our learning process is the pooling step, which is sometimes called the subsampling or downsampling step. This step is mainly for reducing the dimensionality of the output of the convolution step (feature map). The advantage of this pooling step is reducing the size of the feature map while keeping the important information in the newly reduced version.

The following figure shows this step by scanning the image with a 2 by 2 filter and stride 2 while applying the max operation. This kind of pooling operation is called **max pool**:

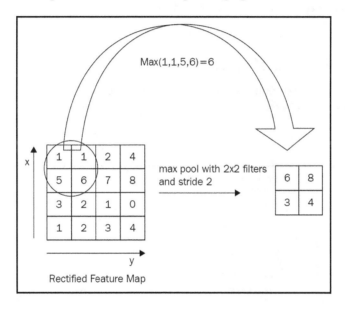

Figure 9.8: An example of a max pooling operation on a rectified feature map (obtained after convolution and ReLU operation) by using a 2 x 2 window (source: http://textminingonline.com/wp-content/uploads/2016/10/max_polling-300x256.png)

We can connect the output of the convolution step to the pooling layer by using the following line of code:

```
pool_layer1 = tf.layers.max_pooling2d(inputs=conv_layer1, pool_size=[2, 2],
strides=2)
```

The pooling layer receives the input from the convolution step with the following shape:

```
[batch_size, image_width, image_height, channels]
```

For example, in our digit classification task, the input to the pooling layer will have the following shape:

```
[batch_size, 28, 28, 20]
```

The output of the pooling operation will have the following shape:

```
[batch_size, 14, 14, 20]
```

In this example, we have reduced the size of the output of the convolution step by 50%. This step is very useful because it keeps only the important information and it also reduces the model's complexity and hence avoids overfitting.

Fully connected layer

After stacking up a bunch of convolution and pooling steps, we follow them with a fully connected layer where we feed the extracted high-level features that we got from the input image to this fully connected layer to use them and do the actual classification based on these features:

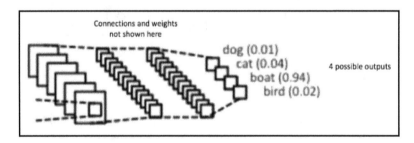

Figure 9.9: Fully connected layer -each node is connected to every other node in the adjacent layer

For example, in the case of the digit classification task, we can follow the convolution and pooling step with a fully connected layer that has 1,024 neurons and ReLU activation to perform the actual classification. This fully connected layer accepts the input in the following format:

```
[batch_size, features]
```

So, we need to reshape or flatten our input feature map from `pool_layer2` to match this format. We can use the following line of code to reshape the output:

```
pool1_flat = tf.reshape(pool_layer1, [-1, 14 * 14 * 20])
```

In this reshape function, we have used −1 to indicate that the batch size will be dynamically determined and each example from the `pool_layer1` output will have a width of 14 and a height of 14 with 20 channels each.

So the final output of this reshape operation will be as follows:

```
[batch_size, 3136]
```

Finally, we can use the `dense()` function of TensorFlow to define our fully connected layer with the required number of neurons (units) and the final activation function:

```
dense_layer = tf.layers.dense(inputs=pool1_flat, units=1024,
activation=tf.nn.relu)
```

Logits layer

Finally, we need the logits layer, which will take the output of the fully connected layer and then produce the raw prediction values. For example, in the case of the digit classification, the output will be a tensor of 10 values, where each value represents the score of one class from 0-9. So, let's define this logit layer for the digit classification example, where we need 10 outputs only, and with linear activation, which is the default for the `dense()` function of TensorFlow:

```
logits_layer = tf.layers.dense(inputs=dense_layer, units=10)
```

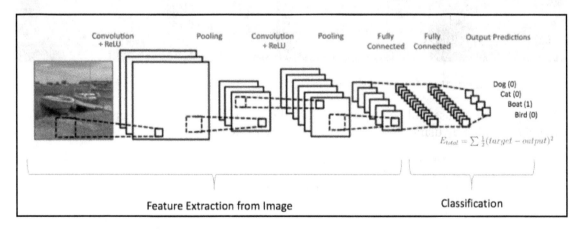

Figure 9.10: Training the ConvNet

The final output of this logits layer will be a tensor of the following shape:

```
[batch_size, 10]
```

As mentioned previously, the logits layer of the model will return the raw predictions our our batch. But we need to convert these values to interpretable format:

- The predicted class for the input sample 0-9.
- The scores or probabilities for each possible class. For example, the probability that the sample is 0, is 1, and so on.

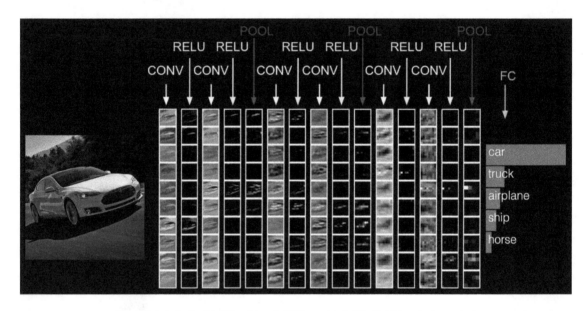

Figure 9.11: A visualization of the different layers of a CNN (source: http://cs231n.github.io/assets/cnn/convnet.jpeg)

So, our predicted class will be the one that has the highest value in the 10 probabilities. We can get this value by using the `argmax` function as follows:

```
tf.argmax(input=logits_layer, axis=1)
```

Remember that the `logits_layer` has the following shape:

```
[batch_size, 10]
```

So, we need to find the max values along our predictions, which is the dimension that has an index of 1.

Finally, we can get our next value, which represents the probabilities of each target class, by applying `softmax` activation to the output of the `logits_layer`, which will squash each value to be between 0 and 1:

```
tf.nn.softmax(logits_layer, name="softmax_tensor")
```

CNN basic example – MNIST digit classification

In this section, we will do a complete example of implementing a CNN for digit classification using the MNIST dataset. We will build a simple model of two convolution layers and fully connected layers.

Let's start off by importing the libraries that will be needed for this implementation:

```
%matplotlib inline
import matplotlib.pyplot as plt
import tensorflow as tf
import numpy as np
from sklearn.metrics import confusion_matrix
import math
```

Next, we will use TensorFlow helper functions to download and preprocess the MNIST dataset as follows:

```
from tensorflow.examples.tutorials.mnist import input_data
mnist_data = input_data.read_data_sets('data/MNIST/', one_hot=True)

Output:
Successfully downloaded train-images-idx3-ubyte.gz 9912422 bytes.
Extracting data/MNIST/train-images-idx3-ubyte.gz
Successfully downloaded train-labels-idx1-ubyte.gz 28881 bytes.
Extracting data/MNIST/train-labels-idx1-ubyte.gz
Successfully downloaded t10k-images-idx3-ubyte.gz 1648877 bytes.
Extracting data/MNIST/t10k-images-idx3-ubyte.gz
Successfully downloaded t10k-labels-idx1-ubyte.gz 4542 bytes.
Extracting data/MNIST/t10k-labels-idx1-ubyte.gz
```

The dataset is split into three disjoint sets: training, validation, and testing. So, let's print the number of images in each set:

```
print("- Number of images in the training
set:\t\t{}".format(len(mnist_data.train.labels)))
print("- Number of images in the test
set:\t\t{}".format(len(mnist_data.test.labels)))
print("- Number of images in the validation
set:\t{}".format(len(mnist_data.validation.labels)))

- Number of images in the training set: 55000
- Number of images in the test set: 10000
- Number of images in the validation set: 5000
```

The actual labels of the images are stored in a one-hot encoding format, so we have an array of 10 values of zeros except for the index of the class that this image represents. For later use, we need to get the class numbers of the dataset as integers:

```
mnist_data.test.cls_integer = np.argmax(mnist_data.test.labels, axis=1)
```

Let's define some known variables to be used later in our implementation:

```
# Default size for the input monocrome images of MNIST
image_size = 28

# Each image is stored as vector of this size.
image_size_flat = image_size * image_size

# The shape of each image
image_shape = (image_size, image_size)

# All the images in the mnist dataset are stored as a monocrome with only 1
channel
num_channels = 1

# Number of classes in the MNIST dataset from 0 till 9 which is 10
num_classes = 10
```

Next, we need to define a helper function to plot some images from the dataset. This helper function will plot the images in a grid of nine subplots:

```
def plot_imgs(imgs, cls_actual, cls_predicted=None):
    assert len(imgs) == len(cls_actual) == 9
    # create a figure with 9 subplots to plot the images.
    fig, axes = plt.subplots(3, 3)
    fig.subplots_adjust(hspace=0.3, wspace=0.3)

    for i, ax in enumerate(axes.flat):
```

```
        # plot the image at the ith index
        ax.imshow(imgs[i].reshape(image_shape), cmap='binary')

        # labeling the images with the actual and predicted classes.
        if cls_predicted is None:
            xlabel = "True: {0}".format(cls_actual[i])
        else:
            xlabel = "True: {0}, Pred: {1}".format(cls_actual[i],
cls_predicted[i])

        # Remove ticks from the plot.
        ax.set_xticks([])
        ax.set_yticks([])
        # Show the classes as the label on the x-axis.
        ax.set_xlabel(xlabel)
    plt.show()
```

Let's plot some images from the test set and see what it looks like:

```
# Visualizing 9 images form the test set.
imgs = mnist_data.test.images[0:9]

# getting the actual classes of these 9 images
cls_actual = mnist_data.test.cls_integer[0:9]

#plotting the images
plot_imgs(imgs=imgs, cls_actual=cls_actual)
```

Here is the output:

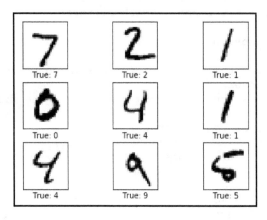

Figure 9.12: A visualization of some examples from the MNIST dataset

Building the model

Now, it's time to build the core of the model. The computational graph includes all the layers we mentioned earlier in this chapter. We'll start by defining some functions that will be used to define variables of a specific shape and randomly initialize them:

```
def new_weights(shape):
    return tf.Variable(tf.truncated_normal(shape, stddev=0.05))

def new_biases(length):
    return tf.Variable(tf.constant(0.05, shape=[length]))
```

Now, let's define the function that will be responsible for creating a new convolution layer based on some input layer, input channels, filter size, number of filters, and whether to use pooling parameters or not:

```
def conv_layer(input,  # the output of the previous layer.
                  input_channels,
                  filter_size,
                  filters,
                  use_pooling=True):  # Use 2x2 max-pooling.

    # preparing the accepted shape of the input Tensor.
    shape = [filter_size, filter_size, input_channels, filters]

    # Create weights which means filters with the given shape.
    filters_weights = new_weights(shape=shape)

    # Create new biases, one for each filter.
    filters_biases = new_biases(length=filters)

    # Calling the conve2d function as we explained above, were the strides
parameter
    # has four values the first one for the image number and the last 1 for
the input image channel
    # the middle ones represents how many pixels the filter should move
with in the x and y axis
    conv_layer = tf.nn.conv2d(input=input,
                        filter=filters_weights,
                        strides=[1, 1, 1, 1],
                        padding='SAME')

    # Adding the biase to the output of the conv_layer.
    conv_layer += filters_biases

    # Use pooling to down-sample the image resolution?
    if use_pooling:
```

```
                 # reduce the output feature map by max_pool layer
                 pool_layer = tf.nn.max_pool(value=conv_layer,
                                     ksize=[1, 2, 2, 1],
                                     strides=[1, 2, 2, 1],
                                     padding='SAME')

        # feeding the output to a ReLU activation function.
        relu_layer = tf.nn.relu(pool_layer)

        # return the final results after applying relu and the filter weights
        return relu_layer, filters_weights
```

As we mentioned previously, the pooling layer produces a 4D tensor. We need to flatten this 4D tensor to a 2D one to be fed to the fully connected layer:

```
def flatten_layer(layer):
    # Get the shape of layer.
    shape = layer.get_shape()

    # We need to flatten the layer which has the shape of The shape
[num_images, image_height, image_width, num_channels]
    # so that it has the shape of [batch_size, num_features] where
number_features is image_height * image_width * num_channels

    number_features = shape[1:4].num_elements()
    # Reshaping that to be fed to the fully connected layer
    flatten_layer = tf.reshape(layer, [-1, number_features])


    # Return both the flattened layer and the number of features.
    return flatten_layer, number_features
```

This function creates a fully connected layer which assumes that the input is a 2D tensor:

```
def fc_layer(input, # the flatten output.
                num_inputs, # Number of inputs from previous layer
                num_outputs, # Number of outputs
                use_relu=True): # Use ReLU on the output to remove
negative values

    # Creating the weights for the neurons of this fc_layer
    fc_weights = new_weights(shape=[num_inputs, num_outputs])
    fc_biases = new_biases(length=num_outputs)

    # Calculate the layer values by doing matrix multiplication of
    # the input values and fc_weights, and then add the fc_bias-values.
    fc_layer = tf.matmul(input, fc_weights) + fc_biases
```

```
# if use RelU parameter is true
if use_relu:
    relu_layer = tf.nn.relu(fc_layer)
    return relu_layer

return fc_layer
```

Before building the network, let's define a placeholder for the input images where the first dimension is `None` to represent an arbitrary number of images:

```
input_values = tf.placeholder(tf.float32, shape=[None, image_size_flat],
name='input_values')
```

As we mentioned previously, the convolution step expects the input images to be in the shape of a 4D tensor. So, we need to reshape the input images to be in the following shape:

```
[num_images, image_height, image_width, num_channels]
```

So, let's reshape the input values to match this format:

```
input_image = tf.reshape(input_values, [-1, image_size, image_size,
num_channels])
```

Next, we need to define another placeholder for the actual class values, which will in one-hot encoding format:

```
y_actual = tf.placeholder(tf.float32, shape=[None, num_classes],
name='y_actual')
```

Also, we need to define a placeholder to hold the integer values of the actual class:

```
y_actual_cls_integer = tf.argmax(y_actual, axis=1)
```

So, let's start off by building the first CNN:

```
conv_layer_1, conv1_weights = \
        conv_layer(input=input_image,
                    input_channels=num_channels,
                    filter_size=filter_size_1,
                    filters=filters_1,
                    use_pooling=True)
```

```
conv_layer_1

<tf.Tensor 'Relu:0' shape=(?, 14, 14, 16) dtype=float32>
```

Next, we will create the second convolution network and feed the output of the first one to it:

```
conv_layer_2, conv2_weights = \
        conv_layer(input=conv_layer_1,
                    input_channels=filters_1,
                    filter_size=filter_size_2,
                    filters=filters_2,
                    use_pooling=True)
```

Also, we need to double-check the shape of the output tensor of the second convolution layer. The shape should be `(?, 7, 7, 36)`, where the ? mark means an arbitrary number of images.

Next, we need to flatten the 4D tensor to match the expected format for the fully connected layer, which is a 2D tensor:

```
flatten_layer, number_features = flatten_layer(conv_layer_2)
```

We need to double-check the shape of the output tensor of the flattened layer:

```
flatten_layer

Output:
<tf.Tensor 'Reshape_1:0' shape=(?, 1764) dtype=float32>
```

Next, we will create a fully connected layer and feed the output of the flattened layer to it. We will also feed the output of the fully connected layer to a ReLU activation function before feeding it to the second fully connected layer:

```
fc_layer_1 = fc_layer(input=flatten_layer,
                        num_inputs=number_features,
                        num_outputs=fc_num_neurons,
                        use_relu=True)
```

```
fc_layer_1
```

```
<tf.Tensor 'Relu_2:0' shape=(?, 128) dtype=float32>
```

Next, we need to add another fully connected layer, which will take the output of the first fully connected layer and produce an array of size 10 for each image that represents the scores for each target class being the correct one:

```
fc_layer_2 = fc_layer(input=fc_layer_1,
                      num_inputs=fc_num_neurons,
                      num_outputs=num_classes,
                      use_relu=False)
```

```
fc_layer_2
```

```
Output:
<tf.Tensor 'add_3:0' shape=(?, 10) dtype=float32>
```

Next, we'll normalize these scores from the second fully connected layer and feed it to a `softmax` activation function, which will squash the values to be between 0 and 1:

```
y_predicted = tf.nn.softmax(fc_layer_2)
```

Then, we need to choose the target class that has the highest probability by using the `argmax` function of TensorFlow:

```
y_predicted_cls_integer = tf.argmax(y_predicted, axis=1)
```

Cost function

Next, we need to define our performance measure, which is the cross-entropy. The value of the cross-entropy will be 0 if the predicted class is correct:

```
cross_entropy = tf.nn.softmax_cross_entropy_with_logits(logits=fc_layer_2,
                                                        labels=y_actual)
```

Next, we need to average all the cross-entropy values that we got from the previous step to be able to get a single performance measure over the test set:

```
model_cost = tf.reduce_mean(cross_entropy)
```

Now, we have a cost function that needs to be optimized/minimized, so we will be using `AdamOptimizer`, which is an optimization method like gradient descent but a bit more advanced:

```
model_optimizer =
tf.train.AdamOptimizer(learning_rate=1e-4).minimize(model_cost)
```

Performance measures

For showing the output, let's define a variable to check whether the predicted class is equal to the true one:

```
model_correct_prediction = tf.equal(y_predicted_cls_integer,
y_actual_cls_integer)
```

Calculate the model accuracy by casting the boolean values then averaging them to sum the correctly classified ones:

```
model_accuracy = tf.reduce_mean(tf.cast(model_correct_prediction,
tf.float32))
```

Model training

Let's kick off the training process by creating a session variable that will be responsible for executing the computational graph that we defined earlier:

```
session = tf.Session()
```

Also, we need to initialize the variables that we have defined so far:

```
session.run(tf.global_variables_initializer())
```

We are going to feed the images in batches to avoid an out-of-memory error:

```
train_batch_size = 64
```

Before kicking the training process, we are going to define a helper function that will perform the optimization process by iterating through the training batches:

```
# number of optimization iterations performed so far
total_iterations = 0

def optimize(num_iterations):
    # Update globally the total number of iterations performed so far.
```

```
    global total_iterations

    for i in range(total_iterations,
                total_iterations + num_iterations):

        # Generating a random batch for the training process
        # input_batch now contains a bunch of images from the training set
and
        # y_actual_batch are the actual labels for the images in the input
batch.
        input_batch, y_actual_batch =
mnist_data.train.next_batch(train_batch_size)

        # Putting the previous values in a dict format for Tensorflow to
automatically assign them to the input
        # placeholders that we defined above
        feed_dict = {input_values: input_batch,
                        y_actual: y_actual_batch}

        # Next up, we run the model optimizer on this batch of images
        session.run(model_optimizer, feed_dict=feed_dict)

        # Print the training status every 100 iterations.
        if i % 100 == 0:
            # measuring the accuracy over the training set.
            acc_training_set = session.run(model_accuracy,
feed_dict=feed_dict)
            #Printing the accuracy over the training set
            print("Iteration: {0:>6}, Accuracy Over the training set:
{1:>6.1%}".format(i + 1, acc_training_set))

    # Update the number of iterations performed so far
    total_iterations += num_iterations
```

And we'll define some helper functions to help us visualize the results of the model and to see which images are misclassified by the model:

```
def plot_errors(cls_predicted, correct):
    # cls_predicted is an array of the predicted class number of each image
in the test set.


    # Extracting the incorrect images.
    incorrect = (correct == False)
    # Get the images from the test-set that have been
    # incorrectly classified.
    images = mnist_data.test.images[incorrect]
    # Get the predicted classes for those incorrect images.
```

```
    cls_pred = cls_predicted[incorrect]

    # Get the actual classes for those incorrect images.
    cls_true = mnist_data.test.cls_integer[incorrect]
    # Plot 9 of these images
    plot_imgs(imgs=imgs[0:9],
                cls_actual=cls_actual[0:9],
                cls_predicted=cls_predicted[0:9])
```

We can also plot the confusion matrix of the predicted results compared to the actual true classes:

```
def plot_confusionMatrix(cls_predicted):

    # cls_predicted is an array of the predicted class number of each image in
    the test set.

    # Get the actual classes for the test-set.
    cls_actual = mnist_data.test.cls_integer

    # Generate the confusion matrix using sklearn.
    conf_matrix = confusion_matrix(y_true=cls_actual,
    y_pred=cls_predicted)

    # Print the matrix.
    print(conf_matrix)

    # visualizing the confusion matrix.
    plt.matshow(conf_matrix)

    plt.colorbar()
    tick_marks = np.arange(num_classes)
    plt.xticks(tick_marks, range(num_classes))
    plt.yticks(tick_marks, range(num_classes))
    plt.xlabel('Predicted class')
    plt.ylabel('True class')

    # Showing the plot
    plt.show()
```

Finally, we are going to define a helper function to help us measure the accuracy of the trained model over the test set:

```
# measuring the accuracy of the trained model over the test set by
splitting it into small batches
test_batch_size = 256

def test_accuracy(show_errors=False,
```

```
                          show_confusionMatrix=False):

    #number of test images
    number_test = len(mnist_data.test.images)

    # define an array of zeros for the predicted classes of the test set
which
    # will be measured in mini batches and stored it.
    cls_predicted = np.zeros(shape=number_test, dtype=np.int)

    # measuring the predicted classes for the testing batches.

    # Starting by the batch at index 0.
    i = 0

    while i < number_test:
        # The ending index for the next batch to be processed is j.
        j = min(i + test_batch_size, number_test)

        # Getting all the images form the test set between the start and
end indices
        input_images = mnist_data.test.images[i:j, :]

        # Get the acutal labels for those images.
        actual_labels = mnist_data.test.labels[i:j, :]

        # Create a feed-dict with the corresponding values for the input
placeholder values
        feed_dict = {input_values: input_images,
                     y_actual: actual_labels}

        cls_predicted[i:j] = session.run(y_predicted_cls_integer,
feed_dict=feed_dict)

        # Setting the start of the next batch to be the end of the one that
we just processed j
        i = j

    # Get the actual class numbers of the test images.
    cls_actual = mnist_data.test.cls_integer

    # Check if the model predictions are correct or not
    correct = (cls_actual == cls_predicted)

    # Summing up the correct examples
    correct_number_images = correct.sum()

    # measuring the accuracy by dividing the correclty classified ones with
```

```
total number of images in the test set.
    testset_accuracy = float(correct_number_images) / number_test

    # showing the accuracy.
    print("Accuracy on Test-Set: {0:.1%} ({1} /
{2})".format(testset_accuracy, correct_number_images, number_test))

    # showing some examples form the incorrect ones.
    if show_errors:
        print("Example errors:")
        plot_errors(cls_predicted=cls_predicted, correct=correct)

    # Showing the confusion matrix of the test set predictions
    if show_confusionMatrix:
        print("Confusion Matrix:")
        plot_confusionMatrix(cls_predicted=cls_predicted)
```

Let's print the accuracy of the created model over the test set without doing any optimization:

```
test_accuracy()

Output:
Accuracy on Test-Set: 4.1% (410 / 10000)
```

Let's get a sense of the optimization process actually enhancing the model capability to classify images to their correct class by running the optimization process for one iteration:

```
optimize(num_iterations=1)
Output:
Iteration: 1, Accuracy Over the training set: 4.7%
test_accuracy()
Output
Accuracy on Test-Set: 4.4% (437 / 10000)
```

Now, let's get down to business and kick off a long optimization process of 10,000 iterations:

```
optimize(num_iterations=9999) #We have already performed 1 iteration.
```

At the end of the output, you should be getting something very close to the following output:

```
Iteration: 7301, Accuracy Over the training set: 96.9%
Iteration: 7401, Accuracy Over the training set: 100.0%
Iteration: 7501, Accuracy Over the training set: 98.4%
Iteration: 7601, Accuracy Over the training set: 98.4%
Iteration: 7701, Accuracy Over the training set: 96.9%
Iteration: 7801, Accuracy Over the training set: 96.9%
Iteration: 7901, Accuracy Over the training set: 100.0%
Iteration: 8001, Accuracy Over the training set: 98.4%
Iteration: 8101, Accuracy Over the training set: 96.9%
Iteration: 8201, Accuracy Over the training set: 100.0%
Iteration: 8301, Accuracy Over the training set: 98.4%
Iteration: 8401, Accuracy Over the training set: 98.4%
Iteration: 8501, Accuracy Over the training set: 96.9%
Iteration: 8601, Accuracy Over the training set: 100.0%
Iteration: 8701, Accuracy Over the training set: 98.4%
Iteration: 8801, Accuracy Over the training set: 100.0%
Iteration: 8901, Accuracy Over the training set: 98.4%
Iteration: 9001, Accuracy Over the training set: 100.0%
Iteration: 9101, Accuracy Over the training set: 96.9%
Iteration: 9201, Accuracy Over the training set: 98.4%
Iteration: 9301, Accuracy Over the training set: 98.4%
Iteration: 9401, Accuracy Over the training set: 100.0%
Iteration: 9501, Accuracy Over the training set: 100.0%
Iteration: 9601, Accuracy Over the training set: 98.4%
Iteration: 9701, Accuracy Over the training set: 100.0%
Iteration: 9801, Accuracy Over the training set: 100.0%
Iteration: 9901, Accuracy Over the training set: 100.0%
Iteration: 10001, Accuracy Over the training set: 98.4%
```

Now, let's check how the model will generalize over the test:

```
test_accuracy(show_errors=True,
                    show_confusionMatrix=True)

Output:
Accuracy on Test-Set: 92.8% (9281 / 10000)
Example errors:
```

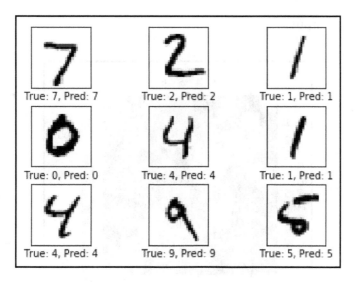

Figure 9.13: Accuracy over the test

```
Confusion Matrix:
[[ 971    0    2    2    0    4    0    1    0    0]
 [   0 1110    4    2    1    2    3    0   13    0]
 [  12    2  949   15   16    3    4   17   14    0]
 [   5    3   14  932    0   34    0   13    6    3]
 [   1    2    3    0  931    1    8    2    3   31]
 [  12    1    4   13    3  852    2    1    3    1]
 [  21    4    5    2   18   34  871    1    2    0]
 [   1   10   26    5    5    0    0  943    2   36]
 [  16    5   10   27   16   48    5   13  815   19]
 [  12    5    5   11   38   10    0   18    3  907]]
```

The following is the output:

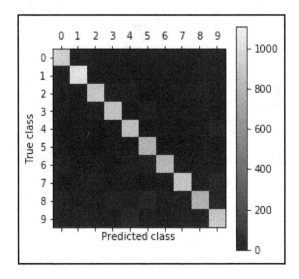

Figure 9.14: Confusion matrix of the test set.

It was interesting that we actually got almost 93% accuracy over the test while using a basic convolution network. This implementation and the results show you what a simple convolution network can do.

Summary

In this chapter, we have covered the intuition and the technical details of how CNNs work. we also had a look at how to implement a basic architecture of a CNN in TensorFlow.

In the next chapter we'll demonstrate more advanced architectures that could be used for detecting objects in one of the image datasets widely used by data scientists. We'll also see the beauty of CNNs and how they come to mimic human understanding of objects by first realizing the basic features of objects and then building up more advanced semantic features on them to come up with a classification for them. Although this process happens very quickly in our minds, it is what actually happens when we recognize objects.

8
Object Detection – CIFAR-10 Example

After covering the basics and the intuition/motivation behind **Convolution Neural Networks** (**CNNs**), we are going to demonstrate this on one of the most popular datasets available for object detection. We'll also see how the initial layers of the CNN get very basic features about our objects, but the final convolutional layers will get more semantic-level features that are built up from those basic features in the first layers.

The following topics will be covered in this chapter:

- Object detection
- CIFAR-10 object detection in mages—model building and training

Object detection

Wikipedia states that:

> "Object detection – technology in the field of computer vision for finding and identifying objects in an image or video sequence. Humans recognize a multitude of objects in images with little effort, despite the fact that the image of the objects may vary somewhat in different view points, in many different sizes and scales or even when they are translated or rotated. Objects can even be recognized when they are partially obstructed from view. This task is still a challenge for computer vision systems. Many approaches to the task have been implemented over multiple decades."

Image analysis is one of the most prominent fields in deep learning. Images are easy to generate and handle, and they are exactly the right type of data for machine learning: easy to understand for human beings, but difficult for computers. Not surprisingly, image analysis played a key role in the history of deep neural networks.

Figure 11.1: Examples of detecting objects. Source: B. C. Russell, A. Torralba, C. Liu, R. Fergus, W. T. Freeman, Object Detection by Scene Alignment, Advances in Neural Information Processing Systems, 2007, at: http://bryanrussell.org/papers/nipsDetectionBySceneAlignment07.pdf

With the rise of autonomous cars, facial detection, smart video surveillance, and people-counting solutions, fast and accurate object detection systems are in a big demand. These systems include not only object recognition and classification in an image, but can also locate each one of them by drawing appropriate boxes around them. This makes object detection a harder task than its traditional computer vision predecessor, image classification.

In this chapter, we'll look at object detection—finding out which objects are in an image. For example, imagine a self-driving car that needs to detect other cars on the road as in *Figure 11.1*. There are lots of complicated algorithms for object detection. They often require huge datasets, very deep convolutional networks, and long training times.

CIFAR-10 – modeling, building, and training

This example shows how to make a CNN for classifying images in the CIFAR-10 dataset. We'll be using a simple convolution neural network implementation of a couple of convolutions and fully connected layers.

Even though the network architecture is very simple, you will see how well it performs when trying to detect objects in the CIFAR-10 images.

So, let's start off this implementation.

Used packages

We import all the required packages for this implementation:

```
%matplotlib inline
%config InlineBackend.figure_format = 'retina'

from urllib.request import urlretrieve
from os.path import isfile, isdir
from tqdm import tqdm
import tarfile
import numpy as np
import random
import matplotlib.pyplot as plt
from sklearn.preprocessing import LabelBinarizer
from sklearn.preprocessing import OneHotEncoder

import pickle
import tensorflow as tf
```

Loading the CIFAR-10 dataset

In this implementation, we'll use CIFAR-10, which is one of the most widely used datasets for object detection. So, let's start off by defining a helper class to download and extract the CIFAR-10 dataset, if it's not already downloaded:

```
cifar10_batches_dir_path = 'cifar-10-batches-py'

tar_gz_filename = 'cifar-10-python.tar.gz'

class DLProgress(tqdm):
    last_block = 0

    def hook(self, block_num=1, block_size=1, total_size=None):
        self.total = total_size
        self.update((block_num - self.last_block) * block_size)
        self.last_block = block_num

if not isfile(tar_gz_filename):
```

```
    with DLProgress(unit='B', unit_scale=True, miniters=1, desc='CIFAR-10
Python Images Batches') as pbar:
        urlretrieve(
            'https://www.cs.toronto.edu/~kriz/cifar-10-python.tar.gz',
            tar_gz_filename,
            pbar.hook)

if not isdir(cifar10_batches_dir_path):
    with tarfile.open(tar_gz_filename) as tar:
        tar.extractall()
        tar.close()
```

After downloading and extracting the CIFAR-10 dataset, you will find out that it's already split into five batches. CIFAR-10 contains images for 10 categories/classes:

- airplane
- automobile
- bird
- cat
- deer
- dog
- frog
- horse
- ship
- truck

Before we dive into building the core of the network, let's do some data analysis and preprocessing.

Data analysis and preprocessing

We need to analyze the dataset and do some basic preprocessing. So, let's start off by defining some helper functions that will enable us to load a specific batch from the five batches that we have and print some analysis about this batch and its samples:

```
# Defining a helper function for loading a batch of images
def load_batch(cifar10_dataset_dir_path, batch_num):
    with open(cifar10_dataset_dir_path + '/data_batch_' + str(batch_num),
mode='rb') as file:
        batch = pickle.load(file, encoding='latin1')
```

```
    input_features = batch['data'].reshape((len(batch['data']), 3, 32,
32)).transpose(0, 2, 3, 1)
    target_labels = batch['labels']

    return input_features, target_labels
```

Then, we define a function that can help us display the stats of a specific sample from a specific batch:

```
#Defining a function to show the stats for batch ans specific sample
def batch_image_stats(cifar10_dataset_dir_path, batch_num, sample_num):

    batch_nums = list(range(1, 6))

    #checking if the batch_num is a valid batch number
    if batch_num not in batch_nums:
        print('Batch Num is out of Range. You can choose from these Batch
nums: {}'.format(batch_nums))
        return None

    input_features, target_labels = load_batch(cifar10_dataset_dir_path,
batch_num)

    #checking if the sample_num is a valid sample number
    if not (0 <= sample_num < len(input_features)):
        print('{} samples in batch {}. {} is not a valid sample
number.'.format(len(input_features), batch_num, sample_num))
        return None

    print('\nStatistics of batch number {}:'.format(batch_num))
    print('Number of samples in this batch:
{}'.format(len(input_features)))
    print('Per class counts of each Label:
{}'.format(dict(zip(*np.unique(target_labels, return_counts=True)))))

    image = input_features[sample_num]
    label = target_labels[sample_num]
    cifar10_class_names = ['airplane', 'automobile', 'bird', 'cat', 'deer',
'dog', 'frog', 'horse', 'ship', 'truck']

    print('\nSample Image Number {}:'.format(sample_num))
    print('Sample image - Minimum pixel value: {} Maximum pixel value:
{}'.format(image.min(), image.max()))
    print('Sample image - Shape: {}'.format(image.shape))
    print('Sample Label - Label Id: {} Name: {}'.format(label,
cifar10_class_names[label]))
    plt.axis('off')
    plt.imshow(image)
```

Now, we can use this function to play around with our dataset and visualize specific images:

```
# Explore a specific batch and sample from the dataset
batch_num = 3
sample_num = 6
batch_image_stats(cifar10_batches_dir_path, batch_num, sample_num)
```

The output is as follows:

```
Statistics of batch number 3:
Number of samples in this batch: 10000
Per class counts of each Label: {0: 994, 1: 1042, 2: 965, 3: 997, 4: 990,
5: 1029, 6: 978, 7: 1015, 8: 961, 9: 1029}

Sample Image Number 6:
Sample image - Minimum pixel value: 30 Maximum pixel value: 242
Sample image - Shape: (32, 32, 3)
Sample Label - Label Id: 8 Name: ship
```

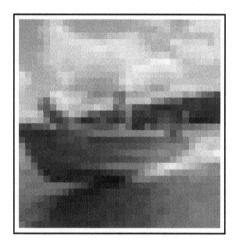

Figure 11.2: Sample image 6 from batch 3

Before going ahead and feeding our dataset to the model, we need to normalize it to the range of zero to one.

Batch normalization optimizes network training. It has been shown to have several benefits:

- **Faster training**: Each training step will be slower because of the extra calculations during the forward pass of the network and the additional hyperparameters to train during backward propagation passes of the network. However, it should converge much more quickly, so training should be faster overall.
- **Higher learning rates**: The gradient descent algorithm mostly requires small learning rates for the network to converge to the loss function's minima. And as the neural networks get deeper, their gradient values get smaller and smaller during backpropagation, so they usually require even more iterations. Using the idea of batch normalization allows us to use much higher learning rates, which further increases the speed at which networks train.
- **Easy to initialize weights**: Weight initialization can be difficult, and it will be even more difficult if we are using deep neural networks. Batch normalization seems to allow us to be much less careful about choosing our initial starting weights.

So, let's proceed by defining a function that will be responsible for normalizing a list of input images so that all the pixel values of these images are between zero and one:

```
#Normalize CIFAR-10 images to be in the range of [0,1]

def normalize_images(images):
    # initial zero ndarray
    normalized_images = np.zeros_like(images.astype(float))
    # The first images index is number of images where the other indices
indicates
    # hieght, width and depth of the image
    num_images = images.shape[0]
    # Computing the minimum and maximum value of the input image to do the
normalization based on them
    maximum_value, minimum_value = images.max(), images.min()
    # Normalize all the pixel values of the images to be from 0 to 1
    for img in range(num_images):
        normalized_images[img,...] = (images[img, ...] -
float(minimum_value)) / float(maximum_value - minimum_value)

    return normalized_images
```

Next up, we need to implement another helper function to encode the labels of the input image. In this function, we will use one-hot encoding of sklearn, where each image label is represented by a vector of zeros except for the class index of the image that this vector represents.

The size of the output vector will be dependent on the number of classes that we have in the dataset, which is 10 classes in the case of CIFAR-10 data:

```
#encoding the input images. Each image will be represented by a vector of
zeros except for the class index of the image
# that this vector represents. The length of this vector depends on number
of classes that we have
# the dataset which is 10 in CIFAR-10

def one_hot_encode(images):
    num_classes = 10
    #use sklearn helper function of OneHotEncoder() to do that
    encoder = OneHotEncoder(num_classes)
    #resize the input images to be 2D
    input_images_resized_to_2d = np.array(images).reshape(-1,1)
    one_hot_encoded_targets =
encoder.fit_transform(input_images_resized_to_2d)
    return one_hot_encoded_targets.toarray()
```

Now, it's time to call the preceding helper functions to do the preprocessing and persist the dataset so that we can use it later:

```
def preprocess_persist_data(cifar10_batches_dir_path, normalize_images,
one_hot_encode):
    num_batches = 5
    valid_input_features = []
    valid_target_labels = []

    for batch_ind in range(1, num_batches + 1):
        #Loading batch
        input_features, target_labels =
load_batch(cifar10_batches_dir_path, batch_ind)
        num_validation_images = int(len(input_features) * 0.1)

        # Preprocess the current batch and perisist it for future use
        input_features = normalize_images(input_features[:-
num_validation_images])
        target_labels = one_hot_encode( target_labels[:-
num_validation_images])
        #Persisting the preprocessed batch
        pickle.dump((input_features, target_labels),
open('preprocess_train_batch_' + str(batch_ind) + '.p', 'wb'))

        # Define a subset of the training images to be used for validating
our model
        valid_input_features.extend(input_features[-
num_validation_images:])
```

```
        valid_target_labels.extend(target_labels[-num_validation_images:])

    # Preprocessing and persisting the validationi subset
    input_features = normalize_images( np.array(valid_input_features))
    target_labels = one_hot_encode(np.array(valid_target_labels))
    pickle.dump((input_features, target_labels), open('preprocess_valid.p',
'wb'))

    #Now it's time to preporcess and persist the test batche
    with open(cifar10_batches_dir_path + '/test_batch', mode='rb') as file:
        test_batch = pickle.load(file, encoding='latin1')


    test_input_features =
test_batch['data'].reshape((len(test_batch['data']), 3, 32,
32)).transpose(0, 2, 3, 1)
    test_input_labels = test_batch['labels']

    # Normalizing and encoding the test batch
    input_features = normalize_images( np.array(test_input_features))
    target_labels = one_hot_encode(np.array(test_input_labels))
    pickle.dump((input_features, target_labels), open('preprocess_test.p',
'wb'))
# Calling the helper function above to preprocess and persist the training,
validation, and testing set
preprocess_persist_data(cifar10_batches_dir_path, normalize_images,
one_hot_encode)
```

So, we have the preprocessed data saved to disk.

We also need to load the validation set for running the trained model on it at different epochs of the training process:

```
# Load the Preprocessed Validation data
valid_input_features, valid_input_labels =
pickle.load(open('preprocess_valid.p', mode='rb'))
```

Building the network

It's now time to build the core of our classification application, which is the computational graph of this CNN architecture, but to maximize the benefits of this implementation, we aren't going to use the TensorFlow layers API. Instead, we are going to use the TensorFlow neural network version of it.

So, let's start off by defining the model input placeholders which will input the images, target classes, and the keep probability parameter of the dropout layer (this helps us to reduce the complexity of the architecture by dropping some connections and hence reducing the chances of overfitting):

```
# Defining the model inputs
def images_input(img_shape):
  return tf.placeholder(tf.float32, (None, ) + img_shape,
name="input_images")

def target_input(num_classes):

  target_input = tf.placeholder(tf.int32, (None, num_classes),
name="input_images_target")
  return target_input

#define a function for the dropout layer keep probability
def keep_prob_input():
  return tf.placeholder(tf.float32, name="keep_prob")
```

Next up, we need to use the TensorFlow neural network implementation version to build up our convolution layers with max pooling:

```
# Applying a convolution operation to the input tensor followed by max
pooling
def conv2d_layer(input_tensor, conv_layer_num_outputs, conv_kernel_size,
conv_layer_strides, pool_kernel_size, pool_layer_strides):


  input_depth = input_tensor.get_shape()[3].value
  weight_shape = conv_kernel_size + (input_depth, conv_layer_num_outputs,)


  #Defining layer weights and biases
  weights = tf.Variable(tf.random_normal(weight_shape))
  biases = tf.Variable(tf.random_normal((conv_layer_num_outputs,)))

  #Considering the biase variable
  conv_strides = (1,) + conv_layer_strides + (1,)


  conv_layer = tf.nn.conv2d(input_tensor, weights, strides=conv_strides,
padding='SAME')
  conv_layer = tf.nn.bias_add(conv_layer, biases)

  conv_kernel_size = (1,) + conv_kernel_size + (1,)
```

```
    pool_strides = (1,) + pool_layer_strides + (1,)
    pool_layer = tf.nn.max_pool(conv_layer, ksize=conv_kernel_size,
strides=pool_strides, padding='SAME')
    return pool_layer
```

As you have probably seen in the previous chapter, the output of the max pooling operation is a 4D tensor, which is not compatible with the required input format for the fully connected layers. So, we need to implement a flattened layer to convert the output of the max pooling layer from 4D to 2D tensor:

```
#Flatten the output of max pooling layer to be fing to the fully connected
layer which only accepts the output
# to be in 2D
def flatten_layer(input_tensor):
return tf.contrib.layers.flatten(input_tensor)
```

Next up, we need to define a helper function that will enable us to add a fully connected layer to our architecture:

```
#Define the fully connected layer that will use the flattened output of the
stacked convolution layers
#to do the actuall classification
def fully_connected_layer(input_tensor, num_outputs):
 return tf.layers.dense(input_tensor, num_outputs)
```

Finally, before using these helper functions to create the entire architecture, we need to create another one that will take the output of the fully connected layer and produce 10 real-valued corresponding to the number of classes that we have in the dataset:

```
#Defining the output function
def output_layer(input_tensor, num_outputs):
     return  tf.layers.dense(input_tensor, num_outputs)
```

So, let's go ahead and define the function that will put all these bits and pieces together and create a CNN with three convolution layers. Each one of them is followed by max pooling operations. We'll also have two fully connected layers, where each one of them is followed by a dropout layer to reduce the model complexity and prevent overfitting. Finally, we'll have the output layer to produce 10 real-valued vectors, where each value represents a score for each class being the correct one:

```
def build_convolution_net(image_data, keep_prob):

    # Applying 3 convolution layers followed by max pooling layers
    conv_layer_1 = conv2d_layer(image_data, 32, (3,3), (1,1), (3,3), (3,3))
    conv_layer_2 = conv2d_layer(conv_layer_1, 64, (3,3), (1,1), (3,3), (3,3))
    conv_layer_3 = conv2d_layer(conv_layer_2, 128, (3,3), (1,1), (3,3), (3,3))
```

```
# Flatten the output from 4D to 2D to be fed to the fully connected layer
  flatten_output = flatten_layer(conv_layer_3)

# Applying 2 fully connected layers with drop out
  fully_connected_layer_1 = fully_connected_layer(flatten_output, 64)
  fully_connected_layer_1 = tf.nn.dropout(fully_connected_layer_1,
keep_prob)
  fully_connected_layer_2 = fully_connected_layer(fully_connected_layer_1,
32)
  fully_connected_layer_2 = tf.nn.dropout(fully_connected_layer_2,
keep_prob)

  #Applying the output layer while the output size will be the number of
categories that we have
  #in CIFAR-10 dataset
  output_logits = output_layer(fully_connected_layer_2, 10)

  #returning output
  return output_logits
```

Let's call the preceding helper functions to build the network and define its loss and optimization criteria:

```
#Using the helper function above to build the network

#First off, let's remove all the previous inputs, weights, biases form the
previous runs
tf.reset_default_graph()

# Defining the input placeholders to the convolution neural network
input_images = images_input((32, 32, 3))
input_images_target = target_input(10)
keep_prob = keep_prob_input()

# Building the models
logits_values = build_convolution_net(input_images, keep_prob)

# Name logits Tensor, so that is can be loaded from disk after training
logits_values = tf.identity(logits_values, name='logits')

# defining the model loss
model_cost =
tf.reduce_mean(tf.nn.softmax_cross_entropy_with_logits(logits=logits_values
, labels=input_images_target))

# Defining the model optimizer
model_optimizer = tf.train.AdamOptimizer().minimize(model_cost)
```

```
# Calculating and averaging the model accuracy
correct_prediction = tf.equal(tf.argmax(logits_values, 1),
tf.argmax(input_images_target, 1))
accuracy = tf.reduce_mean(tf.cast(correct_prediction, tf.float32),
name='model_accuracy')
tests.test_conv_net(build_convolution_net)
```

Now that we have built the computational architecture of this network, it's time to kick off the training process and see some results.

Model training

So, let's define a helper function that will make us able to kick off the training process. This function will take the input images, one-hot encoding of the target classes, and the keep probability value as input. Then, it will feed these values to the computational graph and call the model optimizer:

```
#Define a helper function for kicking off the training process
def train(session, model_optimizer, keep_probability, in_feature_batch,
target_batch):
session.run(model_optimizer, feed_dict={input_images: in_feature_batch,
input_images_target: target_batch, keep_prob: keep_probability})
```

We'll need to validate our model during different time steps in the training process, so we are going to define a helper function that will print out the accuracy of the model on the validation set:

```
#Defining a helper funcitno for print information about the model accuracy
and it's validation accuracy as well
def print_model_stats(session, input_feature_batch, target_label_batch,
model_cost, model_accuracy):
    validation_loss = session.run(model_cost, feed_dict={input_images:
input_feature_batch, input_images_target: target_label_batch, keep_prob:
1.0})
    validation_accuracy = session.run(model_accuracy,
feed_dict={input_images: input_feature_batch, input_images_target:
target_label_batch, keep_prob: 1.0})
    print("Valid Loss: %f" %(validation_loss))
    print("Valid accuracy: %f" % (validation_accuracy))
```

Let's also define the model hyperparameters, which we can use to tune the model for better performance:

```
# Model Hyperparameters
num_epochs = 100
```

```
batch_size = 128
keep_probability = 0.5
```

Now, let's kick off the training process, but only for a single batch of the CIFAR-10 dataset, and see what the model accuracy based on this batch is.

Before that, however, we are going to define a helper function that will load a batch training and also separate the input images from the target classes:

```
# Splitting the dataset features and labels to batches
def batch_split_features_labels(input_features, target_labels,
train_batch_size):
    for start in range(0, len(input_features), train_batch_size):
        end = min(start + train_batch_size, len(input_features))
        yield input_features[start:end], target_labels[start:end]

#Loading the persisted preprocessed training batches
def load_preprocess_training_batch(batch_id, batch_size):
    filename = 'preprocess_train_batch_' + str(batch_id) + '.p'
    input_features, target_labels = pickle.load(open(filename, mode='rb'))

    # Returning the training images in batches according to the batch size
defined above
    return batch_split_features_labels(input_features, target_labels,
train_batch_size)
```

Now, let's start the training process for one batch:

```
print('Training on only a Single Batch from the CIFAR-10 Dataset...')
with tf.Session() as sess:

 # Initializing the variables
 sess.run(tf.global_variables_initializer())

 # Training cycle
 for epoch in range(num_epochs):
 batch_ind = 1

 for batch_features, batch_labels in
load_preprocess_training_batch(batch_ind, batch_size):
 train(sess, model_optimizer, keep_probability, batch_features,
batch_labels)

 print('Epoch number {:>2}, CIFAR-10 Batch Number {}: '.format(epoch + 1,
batch_ind), end='')
 print_model_stats(sess, batch_features, batch_labels, model_cost,
accuracy)
```

```
Output:
    .
    .
    .
Epoch number 85, CIFAR-10 Batch Number 1: Valid Loss: 1.490792
Valid accuracy: 0.550000
Epoch number 86, CIFAR-10 Batch Number 1: Valid Loss: 1.487118
Valid accuracy: 0.525000
Epoch number 87, CIFAR-10 Batch Number 1: Valid Loss: 1.309082
Valid accuracy: 0.575000
Epoch number 88, CIFAR-10 Batch Number 1: Valid Loss: 1.446488
Valid accuracy: 0.475000
Epoch number 89, CIFAR-10 Batch Number 1: Valid Loss: 1.430939
Valid accuracy: 0.550000
Epoch number 90, CIFAR-10 Batch Number 1: Valid Loss: 1.484480
Valid accuracy: 0.525000
Epoch number 91, CIFAR-10 Batch Number 1: Valid Loss: 1.345774
Valid accuracy: 0.575000
Epoch number 92, CIFAR-10 Batch Number 1: Valid Loss: 1.425942
Valid accuracy: 0.575000

Epoch number 93, CIFAR-10 Batch Number 1: Valid Loss: 1.451115
Valid accuracy: 0.550000
Epoch number 94, CIFAR-10 Batch Number 1: Valid Loss: 1.368719
Valid accuracy: 0.600000
Epoch number 95, CIFAR-10 Batch Number 1: Valid Loss: 1.336483
Valid accuracy: 0.600000
Epoch number 96, CIFAR-10 Batch Number 1: Valid Loss: 1.383425
Valid accuracy: 0.575000
Epoch number 97, CIFAR-10 Batch Number 1: Valid Loss: 1.378877
Valid accuracy: 0.625000
Epoch number 98, CIFAR-10 Batch Number 1: Valid Loss: 1.343391
Valid accuracy: 0.600000
Epoch number 99, CIFAR-10 Batch Number 1: Valid Loss: 1.319342
Valid accuracy: 0.625000
Epoch number 100, CIFAR-10 Batch Number 1: Valid Loss: 1.340849
Valid accuracy: 0.525000
```

As you can see, the validation accuracy is not that good while training only on a single batch. Let's see how the validation accuracy is going to change based on only a full training process of the model:

```
model_save_path = './cifar-10_classification'

with tf.Session() as sess:
  # Initializing the variables
  sess.run(tf.global_variables_initializer())
```

```
# Training cycle
for epoch in range(num_epochs):

# iterate through the batches
num_batches = 5

for batch_ind in range(1, num_batches + 1):
for batch_features, batch_labels in
load_preprocess_training_batch(batch_ind, batch_size):
train(sess, model_optimizer, keep_probability, batch_features,
batch_labels)

print('Epoch number{:>2}, CIFAR-10 Batch Number {}: '.format(epoch + 1,
batch_ind), end='')
print_model_stats(sess, batch_features, batch_labels, model_cost,
accuracy)

# Save the trained Model
saver = tf.train.Saver()
save_path = saver.save(sess, model_save_path)
```

Output:
.

.

.
```
Epoch number94, CIFAR-10 Batch Number 5: Valid Loss: 0.316593
Valid accuracy: 0.925000
Epoch number95, CIFAR-10 Batch Number 1: Valid Loss: 0.285429
Valid accuracy: 0.925000
Epoch number95, CIFAR-10 Batch Number 2: Valid Loss: 0.347411
Valid accuracy: 0.825000
Epoch number95, CIFAR-10 Batch Number 3: Valid Loss: 0.232483
Valid accuracy: 0.950000
Epoch number95, CIFAR-10 Batch Number 4: Valid Loss: 0.294707
Valid accuracy: 0.900000
Epoch number95, CIFAR-10 Batch Number 5: Valid Loss: 0.299490
Valid accuracy: 0.975000
Epoch number96, CIFAR-10 Batch Number 1: Valid Loss: 0.302191
Valid accuracy: 0.950000
Epoch number96, CIFAR-10 Batch Number 2: Valid Loss: 0.347043
Valid accuracy: 0.750000
Epoch number96, CIFAR-10 Batch Number 3: Valid Loss: 0.252851
Valid accuracy: 0.875000
Epoch number96, CIFAR-10 Batch Number 4: Valid Loss: 0.291433
Valid accuracy: 0.950000
Epoch number96, CIFAR-10 Batch Number 5: Valid Loss: 0.286192
Valid accuracy: 0.950000
Epoch number97, CIFAR-10 Batch Number 1: Valid Loss: 0.277105
```

```
Valid accuracy: 0.950000
Epoch number97, CIFAR-10 Batch Number 2: Valid Loss: 0.305842
Valid accuracy: 0.850000
Epoch number97, CIFAR-10 Batch Number 3: Valid Loss: 0.215272
Valid accuracy: 0.950000
Epoch number97, CIFAR-10 Batch Number 4: Valid Loss: 0.313761
Valid accuracy: 0.925000
Epoch number97, CIFAR-10 Batch Number 5: Valid Loss: 0.313503
Valid accuracy: 0.925000
Epoch number98, CIFAR-10 Batch Number 1: Valid Loss: 0.265828
Valid accuracy: 0.925000
Epoch number98, CIFAR-10 Batch Number 2: Valid Loss: 0.308948
Valid accuracy: 0.800000
Epoch number98, CIFAR-10 Batch Number 3: Valid Loss: 0.232083
Valid accuracy: 0.950000
Epoch number98, CIFAR-10 Batch Number 4: Valid Loss: 0.298826
Valid accuracy: 0.925000
Epoch number98, CIFAR-10 Batch Number 5: Valid Loss: 0.297230
Valid accuracy: 0.950000
Epoch number99, CIFAR-10 Batch Number 1: Valid Loss: 0.304203
Valid accuracy: 0.900000
Epoch number99, CIFAR-10 Batch Number 2: Valid Loss: 0.308775
Valid accuracy: 0.825000
Epoch number99, CIFAR-10 Batch Number 3: Valid Loss: 0.225072
Valid accuracy: 0.925000
Epoch number99, CIFAR-10 Batch Number 4: Valid Loss: 0.263737
Valid accuracy: 0.925000
Epoch number99, CIFAR-10 Batch Number 5: Valid Loss: 0.278601
Valid accuracy: 0.950000
Epoch number100, CIFAR-10 Batch Number 1: Valid Loss: 0.293509
Valid accuracy: 0.950000
Epoch number100, CIFAR-10 Batch Number 2: Valid Loss: 0.303817
Valid accuracy: 0.875000
Epoch number100, CIFAR-10 Batch Number 3: Valid Loss: 0.244428
Valid accuracy: 0.900000
Epoch number100, CIFAR-10 Batch Number 4: Valid Loss: 0.280712
Valid accuracy: 0.925000
Epoch number100, CIFAR-10 Batch Number 5: Valid Loss: 0.278625
Valid accuracy: 0.950000
```

Testing the model

Let's test the trained model against the test set part of the CIFAR-10 dataset. First, we are going to define a helper function that will help us to visualize the predictions of some sample images and their corresponding true labels:

```
#A helper function to visualize some samples and their corresponding
predictions
def display_samples_predictions(input_features, target_labels,
samples_predictions):

 num_classes = 10

 cifar10_class_names = ['airplane', 'automobile', 'bird', 'cat', 'deer',
'dog', 'frog', 'horse', 'ship', 'truck']

 label_binarizer = LabelBinarizer()
 label_binarizer.fit(range(num_classes))
 label_inds = label_binarizer.inverse_transform(np.array(target_labels))

 fig, axies = plt.subplots(nrows=4, ncols=2)
 fig.tight_layout()
 fig.suptitle('Softmax Predictions', fontsize=20, y=1.1)

 num_predictions = 4
 margin = 0.05
 ind = np.arange(num_predictions)
 width = (1. - 2. * margin) / num_predictions

 for image_ind, (feature, label_ind, prediction_indicies, prediction_values)
 in enumerate(zip(input_features, label_inds, samples_predictions.indices,
samples_predictions.values)):
  prediction_names = [cifar10_class_names[pred_i] for pred_i in
prediction_indicies]
  correct_name = cifar10_class_names[label_ind]

 axies[image_ind][0].imshow(feature)
  axies[image_ind][0].set_title(correct_name)
  axies[image_ind][0].set_axis_off()

 axies[image_ind][1].barh(ind + margin, prediction_values[::-1], width)
  axies[image_ind][1].set_yticks(ind + margin)
  axies[image_ind][1].set_yticklabels(prediction_names[::-1])
  axies[image_ind][1].set_xticks([0, 0.5, 1.0])
```

Now, let's restore the trained model and test it against the test set:

```
test_batch_size = 64
save_model_path = './cifar-10_classification'
#Number of images to visualize
num_samples = 4

#Number of top predictions
top_n_predictions = 4

#Defining a helper function for testing the trained model
def test_classification_model():

 input_test_features, target_test_labels =
pickle.load(open('preprocess_test.p', mode='rb'))
 loaded_graph = tf.Graph()
with tf.Session(graph=loaded_graph) as sess:

 # loading the trained model
 model = tf.train.import_meta_graph(save_model_path + '.meta')
 model.restore(sess, save_model_path)

# Getting some input and output Tensors from loaded model
 model_input_values = loaded_graph.get_tensor_by_name('input_images:0')
 model_target = loaded_graph.get_tensor_by_name('input_images_target:0')
 model_keep_prob = loaded_graph.get_tensor_by_name('keep_prob:0')
 model_logits = loaded_graph.get_tensor_by_name('logits:0')
 model_accuracy = loaded_graph.get_tensor_by_name('model_accuracy:0')

 # Testing the trained model on the test set batches
 test_batch_accuracy_total = 0
 test_batch_count = 0

 for input_test_feature_batch, input_test_label_batch in
batch_split_features_labels(input_test_features, target_test_labels,
test_batch_size):
 test_batch_accuracy_total += sess.run(
 model_accuracy,
 feed_dict={model_input_values: input_test_feature_batch, model_target:
input_test_label_batch, model_keep_prob: 1.0})
 test_batch_count += 1

print('Test set accuracy:
{}\n'.format(test_batch_accuracy_total/test_batch_count))

# print some random images and their corresponding predictions from the
test set results
 random_input_test_features, random_test_target_labels =
```

```
tuple(zip(*random.sample(list(zip(input_test_features,
target_test_labels)), num_samples)))

 random_test_predictions = sess.run(
 tf.nn.top_k(tf.nn.softmax(model_logits), top_n_predictions),
 feed_dict={model_input_values: random_input_test_features, model_target:
random_test_target_labels, model_keep_prob: 1.0})

 display_samples_predictions(random_input_test_features,
random_test_target_labels, random_test_predictions)

#Calling the function
test_classification_model()
```

```
Output:
INFO:tensorflow:Restoring parameters from ./cifar-10_classification
Test set accuracy: 0.7540007961783439
```

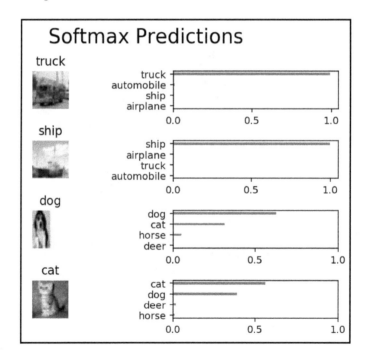

Let's visualize another example to see some errors:

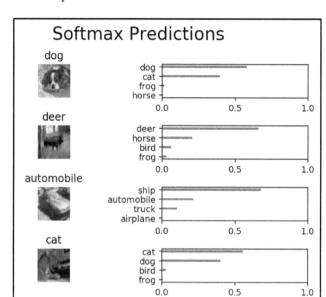

Now, we have a test accuracy of around 75%, which is not bad for a simple CNN like the one we have used.

Summary

This chapter showed us how to make a CNN for classifying images in the CIFAR-10 dataset. The classification accuracy was about 79% - 80% on the test set. The output of the convolutional layers was also plotted, but it was difficult to see how the neural network recognizes and classifies the input images. Better visualization techniques are needed.

Next up, we'll use one of the modern and exciting practice of deep learning, which is transfer learning. Transfer learning allows you to use data-greedy architectures of deep learning with small datasets.

Object Detection – Transfer Learning with CNNs

9

"How individuals transfer in one context to another context that share similar characteristics"

– E. L. Thorndike, R. S. Woodworth (1991)

Transfer learning (TL) is a research problem in data science that is mainly concerned with persisting knowledge acquired during solving a specific task and using this acquired knowledge to solve another different but similar task. In this chapter, we will demonstrate one of the modern practices and common themes used in the field of data science with TL. The idea here is how to get the help from domains with very large datasets to domains that have less dataset size. Finally, we will revisit our object detection example of CIFAR-10 and try to reduce both the training time and performance error via TL.

The following topics will be covered in this chapter:

- Transfer learning
- CIFAR-10 object detection revisited

Transfer learning

Deep learning architectures are data greedy and having a few samples in a training set will not get us the best out of them. TL solves this problem by transferring learned or gained knowledge/representations from solving a task with a large dataset to another different but similar one with a smaller dataset.

TL is not only useful for the case of small training sets, but also we can use it to make the training process faster. Training large deep learning architectures from scratch can sometimes be very slow because we have millions of weights in these architectures that need to be learned. Instead, someone can make use of TL by just fine-tuning a learned weight on a similar problem to the one that he/she's trying to solve.

The intuition behind TL

Let's build up the intuition behind TL by using the following teacher-student analogy. A teacher has many years of experience in the modules that he'she's teaching. On the other side, the students get a compact overview of the topic from the lectures that this teacher gives. So you can say that the teacher is transferring their knowledge in a concise and compact way to the students.

The same analogy of the teacher and students can be applied to our case of transferring knowledge in deep learning, or in neural networks in general. So our model learns some representations from the data, which is represented by the *weights* of the network. These learned representations/features (weights) can be transferred to another different but similar task. This process of transferring the learned weights to another task will reduce the need for huge datasets for deep learning architectures to converge, and it will also reduce the time needed to adapt the model to the new dataset compared to training the model from scratch.

Deep learning is widely used nowadays, but usually most people are using TL while training deep learning architectures; few of them train deep learning architectures from scratch, because most of the time it's rare to have a dataset of sufficient size for deep learning to converge. So it's very common to use a pre-trained model on a large dataset such as `ImageNet`, which has about 1.2 million images, and apply it to your new task. We can use the weights of that pre-trained model as a feature extractor, or we can just initialize our architecture with it and then fine-tune them to your new task. There are three major scenarios for using TL:

- **Use a convolution network as a fixed feature extractor**: In this scenario, you use a pre-trained convolution model on a large dataset such as ImageNet and adapt it to work on your problem. For instance, a pre-trained convolution model on ImageNet will have a fully connected layer with output scores for the 1,000 categories that ImageNet has. So you need to remove this layer because you are not interested anymore in the classes of ImageNet. Then, you treat all other layers as a feature extractor. Once you have extracted the features using the pre-trained model, you can feed these features to any linear classifier, such as the softmax classifier, or even linear SVM.

- **Fine-tune the convolution neural network**: The second scenario involves the first one but with an extra effort to fine-tune the pre-trained weights on your new task using backpropagation. Usually, people keep most of the layers fixed and only fine-tune the top end of the network. Trying to fine-tune the whole network or even most of the layers may result in overfitting. So, you might be interested in fine-tuning only those layers that are concerned with the semantic-level features of the images. The intuition behind leaving the earlier layers fixed is that they contain generic or low-level features that are common across most imaging tasks, such as corners, edges, and so on. Fine-tuning the higher level or the top end layers of the network will be useful if you're introducing new classes that are not present in the original dataset that the model was pre-trained on.

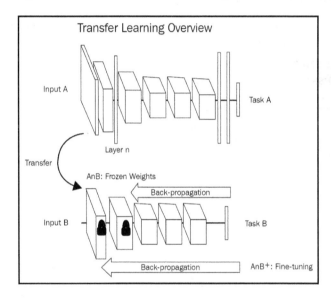

Figure 10.1: Fine-tuning the pre-trained CNN for a new task

- **Pre-trained models**: The third widely used scenario is to download checkpoints that people have made available on the internet. You may go for this scenario if you don't have big computational power to train the model from scratch, so you just initialize the model with the released checkpoints and then do a little fine-tuning.

Differences between traditional machine learning and TL

As you've noticed from the previous section, there's a clear difference between the traditional way we apply machine learning and machine learning that involves TL (as shown in the following diagram). In traditional machine learning, you don't transfer any knowledge or representations to any other task, which is not the case in TL. Sometimes, people use TL in a wrong way, so we are going to mention a few conditions under which you can only use TL to maximize the gains.

The following are the conditions for applying TL:

- Unlike traditional machine learning, the source and target task or domains don't have to come from the same distribution, but they have to be similar
- You can also use TL in case of less training samples or if you don't have the necessary computational power

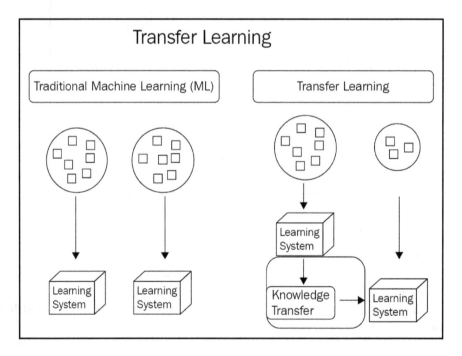

Figure 10.2: Traditional machine learning versus machine learning with TL

CIFAR-10 object detection – revisited

In the previous chapter, we trained a simple **convolution neural network** (**CNN**) model on the CIFAR-10 dataset. Here, we are going to demonstrate the case of using a pre-trained model as a feature extractor while removing the fully connected layer of the pre-trained model, and then we'll feed these extracted features or transferred values to a softmax layer.

The pre-trained model in this implementation will be the inception model, which will be pre-trained on ImageNet. But bear in mind that this implementation builds on the previous two chapters that introduced CNN.

Solution outline

Again, we are going to replace the final fully connected layer of the pre-trained inception model and then use the rest of the inception model as a feature extractor. So, we first feed our raw images in the inception model, which will extract the features from them and then output our so-called transfer values.

After getting the transfer values of the extracted features from the inception model, you might need to save them to your desk because it will take some time if you did it on the fly, so it's useful to persist them to your desk to save you time. In TensorFlow tutorials, they use the term bottleneck values instead of transfer values, but it's just a different name for the exact same thing.

After getting the transfer values or loading them from the desk, we can feed them to any linear classifier that's customized to our new task. Here, we will feed the extracted transfer values to another neural network and then train for the new classes of CIFAR-10.

The following diagram, shows the general solution outline that we will be following:

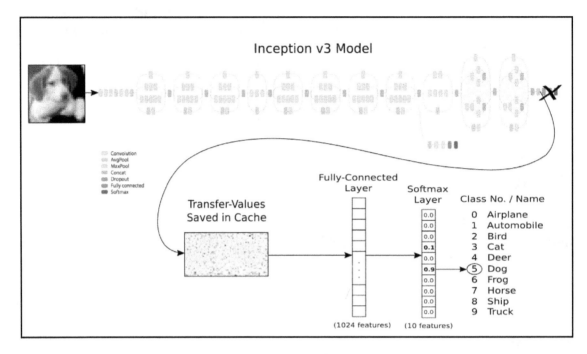

Figure 10.3: The solution outline for an object detection task using the CIFAR-10 dataset with TL

Loading and exploring CIFAR-10

Let's start off by importing the required packages for this implementation:

```
%matplotlib inline
import matplotlib.pyplot as plt
import tensorflow as tf
import numpy as np
import time
from datetime import timedelta
import os

# Importing a helper module for the functions of the Inception model.
import inception
```

Next up, we need to load another helper script that we can use to download the processing CIFAR-10 dataset:

```
import cifar10
#importing number of classes of CIFAR-10
from cifar10 import num_classes
```

If you haven't done this already, you need to set the path for CIFAR-10. This path will be used by the `cifar-10.py` script to persist the dataset:

```
cifar10.data_path = "data/CIFAR-10/"
```

The CIFAR-10 dataset is about 170 MB, the next line checks if the dataset is already downloaded if not it downloads the dataset and store in the previous data_path:

```
cifar10.maybe_download_and_extract</span>()
```

```
Output:
```

```
- Download progress: 100.0%
Download finished. Extracting files.
Done.
```

Let's see the categories that we have in the CIFAR-10 dataset:

```
#Loading the class names of CIFAR-10 dataset
class_names = cifar10.load_class_names()
class_names
```

Output:

```
Loading data: data/CIFAR-10/cifar-10-batches-py/batches.meta
['airplane',
 'automobile',
 'bird',
 'cat',
 'deer',
 'dog',
 'frog',
 'horse',
 'ship',
 'truck']
Load the training-set.
```

This returns `images`, the class-numbers as `integers`, and the class-numbers as one-hot encoded arrays called `labels`:

```
training_images, training_cls_integers, trainig_one_hot_labels =
cifar10.load_training_data()
```

Output:

```
Loading data: data/CIFAR-10/cifar-10-batches-py/data_batch_1
Loading data: data/CIFAR-10/cifar-10-batches-py/data_batch_2
Loading data: data/CIFAR-10/cifar-10-batches-py/data_batch_3
Loading data: data/CIFAR-10/cifar-10-batches-py/data_batch_4
Loading data: data/CIFAR-10/cifar-10-batches-py/data_batch_5
Load the test-set.
```

Now, let's do the same for the testing set by loading the images and their corresponding integer representation of the target classes with their one-hot encoding:

```
#Loading the test images, their class integer, and their corresponding one-
hot encoding
testing_images, testing_cls_integers, testing_one_hot_labels =
cifar10.load_test_data()
```

```
Output:
```

```
Loading data: data/CIFAR-10/cifar-10-batches-py/test_batch
```

Let's have a look at the distribution of the training and testing sets in CIFAR-10:

```
print("-Number of images in the training
set:\t\t{}".format(len(training_images)))
print("-Number of images in the testing
set:\t\t{}".format(len(testing_images)))
```

Output:

```
-Number of images in the training set:          50000
-Number of images in the testing set:           10000
```

Let's define some helper functions that will enable us to explore the dataset. The following helper function plots a set of nine images in a grid:

```
def plot_imgs(imgs, true_class, predicted_class=None):

    assert len(imgs) == len(true_class)

    # Creating a placeholders for 9 subplots
    fig, axes = plt.subplots(3, 3)
```

```
    # Adjustting spacing.
    if predicted_class is None:
        hspace = 0.3
    else:
        hspace = 0.6
    fig.subplots_adjust(hspace=hspace, wspace=0.3)


    for i, ax in enumerate(axes.flat):
        # There may be less than 9 images, ensure it doesn't crash.
        if i < len(imgs):
            # Plot image.
            ax.imshow(imgs[i],
                      interpolation='nearest')

            # Get the actual name of the true class from the class_names
array
            true_class_name = class_names[true_class[i]]

            # Showing labels for the predicted and true classes
            if predicted_class is None:
                xlabel = "True: {0}".format(true_class_name)
            else:
                # Name of the predicted class.
                predicted_class_name = class_names[predicted_class[i]]

                xlabel = "True: {0}\nPred: {1}".format(true_class_name,
predicted_class_name)

            ax.set_xlabel(xlabel)
        # Remove ticks from the plot.
        ax.set_xticks([])
        ax.set_yticks([])
    plt.show()
```

Let's go ahead and visualize some images from the test set along with their corresponding actual class:

```
# get the first 9 images in the test set
imgs = testing_images[0:9]

# Get the integer representation of the true class.
true_class = testing_cls_integers[0:9]

# Plotting the images
plot_imgs(imgs=imgs, true_class=true_class)
```

Output:

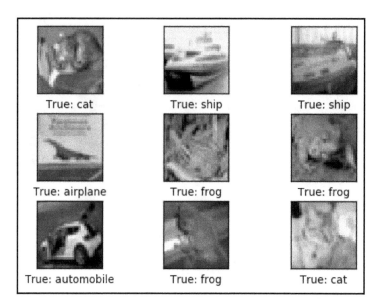

Figure 10.4: The first nine images of the test set

Inception model transfer values

As we mentioned earlier, we will be using the pre-trained inception model on the ImageNet dataset. So, we need to download this pre-trained model from the internet.

Let's start off by defining `data_dir` for the inception model:

```
inception.data_dir = 'inception/'
```

The weights of the pre-trained inception model are about 85 MB. The following line of code will download it if it doesn't exist in the `data_dir` defined previously:

```
inception.maybe_download()

Downloading Inception v3 Model ...
- Download progress: 100%
```

We will load the inception model so that we can use it as a feature extractor for our CIFAR-10 images:

```
# Loading the inception model so that we can inialized it with the pre-
trained weights and customize for our model
inception_model = inception.Inception()
```

As we mentioned previously, calculating the transfer values for the CIFAR-10 dataset will take some time, so we need to cache them for future use. Thankfully, there's a helper function in the `inception` module that can help us do that:

```
from inception import transfer_values_cache
```

Next up, we need to set the file paths for the cached training and testing files:

```
file_path_train = os.path.join(cifar10.data_path,
'inception_cifar10_train.pkl')
file_path_test = os.path.join(cifar10.data_path,
'inception_cifar10_test.pkl')
print("Processing Inception transfer-values for the training images of
Cifar-10 ...")
# First we need to scale the imgs to fit the Inception model requirements
as it requires all pixels to be from 0 to 255,
# while our training examples of the CIFAR-10 pixels are between 0.0 and
1.0
imgs_scaled = training_images * 255.0

# Checking if the transfer-values for our training images are already
calculated and loading them, if not calculate and save them.
transfer_values_training =
transfer_values_cache(cache_path=file_path_train,
                                        images=imgs_scaled,
                                        model=inception_model)
print("Processing Inception transfer-values for the testing images of
Cifar-10 ...")
# First we need to scale the imgs to fit the Inception model requirements
as it requires all pixels to be from 0 to 255,
# while our training examples of the CIFAR-10 pixels are between 0.0 and
1.0
imgs_scaled = testing_images * 255.0
# Checking if the transfer-values for our training images are already
calculated and loading them, if not calcaulate and save them.
transfer_values_testing = transfer_values_cache(cache_path=file_path_test,
                            images=imgs_scaled,
                            model=inception_model)
```

As mentioned before, we have 50,000 images in the training set of the CIFAR-10 dataset. So let's check the shapes of the transfer values of these images. It should be 2,048 for each image in this training set:

```
transfer_values_training.shape
```

Output:

```
(50000, 2048)
```

We need to do the same for the test set:

```
transfer_values_testing.shape
```

Output:

```
(10000, 2048)
```

To intuitively understand how the transfer values look, we are going to define a helper function to enable us to use the plot the transfer values of a specific image from the training or the testing sets:

```
def plot_transferValues(ind):
    print("Original input image:")
    # Plot the image at index ind of the test set.
    plt.imshow(testing_images[ind], interpolation='nearest')
    plt.show()

    print("Transfer values using Inception model:")
    # Visualize the transfer values as an image.
    transferValues_img = transfer_values_testing[ind]
    transferValues_img = transferValues_img.reshape((32, 64))

    # Plotting the transfer values image.
    plt.imshow(transferValues_img, interpolation='nearest', cmap='Reds')
    plt.show()
plot_transferValues(i=16)

Input image:
```

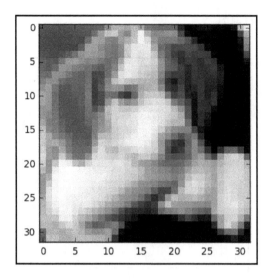

Figure 10.5: Input image

Transfer values for the image using the inception model:

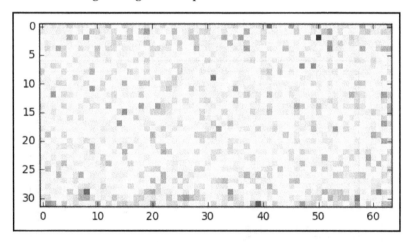

Figure 10.6: Transfer values for the input image in Figure 10.3

```
plot_transferValues(i=17)
```

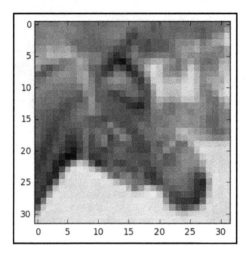

Figure 10.7: Input image

Transfer values for the image using the inception model:

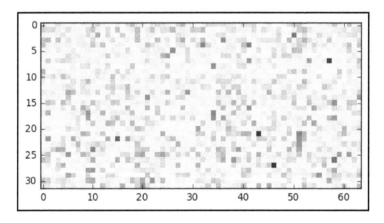

Figure 10.8: Transfer values for the input image in Figure 10.5

Analysis of transfer values

In this section, we will do some analysis of the transferred values that we just got for the training images. The purpose of this analysis is to see whether these transfer values will be enough for classifying the images that we have in CIFAR-10 or not.

We have 2,048 transfer values for each input image. In order to plot these transfer values and do further analysis on them, we can use dimensionality reduction techniques such as **Principal Component Analysis** (**PCA**) from scikit-learn. We'll reduce the transfer values from 2,048 to 2 to be able to visualize it and see if they will be good features for discriminating between different categories of CIFAR-10:

```
from sklearn.decomposition import PCA
```

Next up, we need to create a PCA object wherein the number of components is only 2:

```
pca_obj = PCA(n_components=2)
```

It takes a lot of time to reduce the transfer values from 2,048 to 2, so we are going to subset only 3,000 out of the 5,000 images that we have transfer values for:

```
subset_transferValues = transfer_values_training[0:3000]
```

We need to get the class numbers of these images as well:

```
cls_integers = testing_cls_integers[0:3000]
```

We can double-check our subsetting by printing the shape of the transfer values:

```
subset_transferValues.shape
```

Output:

```
(3000, 2048)
```

Next up, we use our PCA object to reduce the transfer values from 2,048 to just 2:

```
reduced_transferValues = pca_obj.fit_transform(subset_transferValues)
```

Now, let's see the output of the PCA reduction process:

```
reduced_transferValues.shape
```

Output:
```
(3000, 2)
```

After reducing the dimensionality of the transfer values to only 2, let's plot these values:

```
#Importing the color map for plotting each class with different color.
import matplotlib.cm as color_map

def plot_reduced_transferValues(transferValues, cls_integers):
    # Create a color-map with a different color for each class.
    c_map = color_map.rainbow(np.linspace(0.0, 1.0, num_classes))

    # Getting the color for each sample.
    colors = c_map[cls_integers]

    # Getting the x and y values.
    x_val = transferValues[:, 0]
    y_val = transferValues[:, 1]

    # Plot the transfer values in a scatter plot
    plt.scatter(x_val, y_val, color=colors)
    plt.show()
```

Here, we are plotting the reduced transfer values of the subset from the training set. We have 10 classes in CIFAR-10, so we are going to plot their corresponding transfer values with different colors. As you can see from the following graph, the transfer values are grouped according to the corresponding class. The overlap between groups is because the reduction process of PCA can't properly separate the transfer values:

```
plot_reduced_transferValues(reduced_transferValues, cls_integers)
```

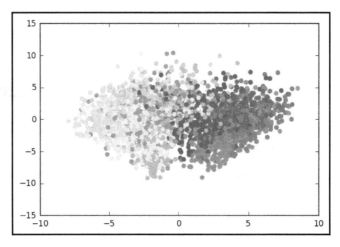

Figure 10.9: Transfer values reduced using PCA

We can do a further analysis on our transfer values using a different dimensionality reduction method called **t-SNE**:

```
from sklearn.manifold import TSNE
```

Again, we'll be reduce our dimensionality of the transfer values, which is 2,048, but this time to 50 values and not 2:

```
pca_obj = PCA(n_components=50)
transferValues_50d = pca_obj.fit_transform(subset_transferValues)
```

Next up, we stack the second dimensionality reduction technique and feed the output of the PCA process to it:

```
tsne_obj = TSNE(n_components=2)
```

Finally, we use the reduced values from the PCA method and apply the t-SNE method to it:

```
reduced_transferValues = tsne_obj.fit_transform(transferValues_50d)
```

And double-check if it has the correct shape:

```
reduced_transferValues.shape
```

Output:

```
(3000, 2)
```

Let's plot the reduced transfer values by the t-SNE method. As you can see in the next image, the t-SNE has been able to do better separation of grouped transfer values than the PCA one.

The takeaway from this analysis is that the extracted transfer values we got by feeding our input images to the pre-trained inception model can be used to separate training images into the 10 classes. This separation won't be 100% accurate because of the small overlap in the following graph, but we can get rid of this overlap by doing some fine-tuning on our pre-trained model:

```
plot_reduced_transferValues(reduced_transferValues, cls_integers)
```

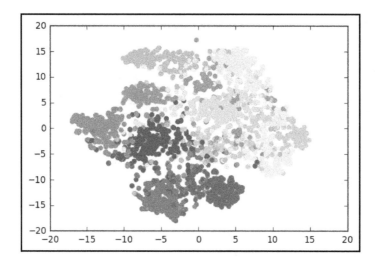

Figure 10.10: Transfer values reduced using t-SNE

Now we have transfer values extracted from our training images and we know that these values will be able to, to some extent, distinguish between the different classes that CIFAR-10 has. Next, we need to build a linear classifier and feed these transfer values to it to do the actual classification.

Model building and training

So, let's start off by specifying the input placeholder variables that will be fed to our neural network model. The shape of the first input variable (which will contain the extracted transfer values) will be [None, transfer_len]. The second placeholder variable will hold the actual class labels of the training set in a one-hot vector format:

```
transferValues_arrLength = inception_model.transfer_len
input_values = tf.placeholder(tf.float32, shape=[None,
transferValues_arrLength], name='input_values')
y_actual = tf.placeholder(tf.float32, shape=[None, num_classes],
name='y_actual')
```

We can also get the corresponding integer value of each class from 1 to 10 by defining another placeholder variable:

```
y_actual_cls = tf.argmax(y_actual, axis=1)
```

Next up, we need to build the actual classification neural network that will take these input placeholders and produce the predicted classes:

```
def new_weights(shape):
    return tf.Variable(tf.truncated_normal(shape, stddev=0.05))

def new_biases(length):
    return tf.Variable(tf.constant(0.05, shape=[length]))

def new_fc_layer(input,          # The previous layer.
                 num_inputs,     # Num. inputs from prev. layer.
                 num_outputs,    # Num. outputs.
                 use_relu=True): # Use Rectified Linear Unit (ReLU)?

    # Create new weights and biases.
    weights = new_weights(shape=[num_inputs, num_outputs])
    biases = new_biases(length=num_outputs)

    # Calculate the layer as the matrix multiplication of
    # the input and weights, and then add the bias-values.
    layer = tf.matmul(input, weights) + biases

    # Use ReLU?
    if use_relu:
        layer = tf.nn.relu(layer)

    return layer

# First fully-connected layer.
layer_fc1 = new_fc_layer(input=input_values,
                         num_inputs=2048,
                         num_outputs=1024,
                         use_relu=True)

# Second fully-connected layer.
layer_fc2 = new_fc_layer(input=layer_fc1,
                         num_inputs=1024,
                         num_outputs=num_classes,
                         use_relu=False)

# Predicted class-label.
y_predicted = tf.nn.softmax(layer_fc2)
```

```
# Cross-entropy for the classification of each image.
cross_entropy = \
    tf.nn.softmax_cross_entropy_with_logits(logits=layer_fc2,
                                            labels=y_actual)

# Loss aka. cost-measure.
# This is the scalar value that must be minimized.
loss = tf.reduce_mean(cross_entropy)
```

Then, we need to define an optimization criteria that will be used during the training of the classifier. In this implementation, we will use `AdamOptimizer`. The output of this classifier will be an array of 10 probability scores, corresponding to the number of classes that we have in the CIFAR-10 dataset. Then, we are going to apply the `argmax` operation over this array to assign the class of the largest score to this input sample:

```
step = tf.Variable(initial_value=0,
                              name='step', trainable=False)
optimizer = tf.train.AdamOptimizer(learning_rate=1e-4).minimize(loss, step)
y_predicted_cls = tf.argmax(y_predicted, axis=1)
#compare the predicted and true classes
correct_prediction = tf.equal(y_predicted_cls, y_actual_cls)
#cast the boolean values to fload
model_accuracy = tf.reduce_mean(tf.cast(correct_prediction, tf.float32))
```

Next up, we need to define a TensorFlow session that will actually execute the graph and then initialize the variables that we defined earlier in this implementation:

```
session = tf.Session()
session.run(tf.global_variables_initializer())
```

In this implementation, we will be using **Stochastic Gradient Descent (SGD)**, so we need to define a function to randomly generate batches of a particular size from our training set of 50,000 images.

Thus, we are going to define a helper function for generating a random batch from the input training set of the transfer values:

```
#defining the size of the train batch
train_batch_size = 64

#defining a function for randomly selecting a batch of images from the
dataset
def select_random_batch():
    # Number of images (transfer-values) in the training-set.
    num_imgs = len(transfer_values_training)

    # Create a random index.
```

```
ind = np.random.choice(num_imgs,
                       size=training_batch_size,
                       replace=False)

# Use the random index to select random x and y-values.
# We use the transfer-values instead of images as x-values.
x_batch = transfer_values_training[ind]
y_batch = trainig_one_hot_labels[ind]

return x_batch, y_batch
```

Next up, we need to define a helper function to do the actual optimization process, which will refine the weights of the network. It will generate a batch at each iteration and optimize the network based on that batch:

```
def optimize(num_iterations):

    for i in range(num_iterations):
        # Selectin a random batch of images for training
        # where the transfer values of the images will be stored in
input_batch
        # and the actual labels of those batch of images will be stored in
y_actual_batch
        input_batch, y_actual_batch = select_random_batch()

        # storing the batch in a dict with the proper names
        # such as the input placeholder variables that we define above.
        feed_dict = {input_values: input_batch,
                     y_actual: y_actual_batch}

        # Now we call the optimizer of this batch of images
        # TensorFlow will automatically feed the values of the dict we
created above
        # to the model input placeholder variables that we defined above.
        i_global, _ = session.run([step, optimizer],
                                  feed_dict=feed_dict)

        # print the accuracy every 100 steps.
        if (i_global % 100 == 0) or (i == num_iterations - 1):
            # Calculate the accuracy on the training-batch.
            batch_accuracy = session.run(model_accuracy,
                                         feed_dict=feed_dict)

            msg = "Step: {0:>6}, Training Accuracy: {1:>6.1%}"
            print(msg.format(i_global, batch_accuracy))
```

We are going to define some helper functions to show the results of the previous neural network and show the confusion matrix of the predicted results as well:

```
def plot_errors(cls_predicted, cls_correct):
    # cls_predicted is an array of the predicted class-number for
    # all images in the test-set.

    # cls_correct is an array with boolean values to indicate
    # whether is the model predicted the correct class or not.

    # Negate the boolean array.
    incorrect = (cls_correct == False)
    # Get the images from the test-set that have been
    # incorrectly classified.
    incorrectly_classified_images = testing_images[incorrect]
    # Get the predicted classes for those images.
    cls_predicted = cls_predicted[incorrect]

    # Get the true classes for those images.
    true_class = testing_cls_integers[incorrect]

    n = min(9, len(incorrectly_classified_images))
    # Plot the first n images.
    plot_imgs(imgs=incorrectly_classified_images[0:n],
              true_class=true_class[0:n],
              predicted_class=cls_predicted[0:n])
```

Next, we need to define the helper function for plotting the confusion matrix:

```
from sklearn.metrics import confusion_matrix

def plot_confusionMatrix(cls_predicted):

    # cls_predicted array of all the predicted
    # classes numbers in the test.

    # Call the confucion matrix of sklearn
    cm = confusion_matrix(y_true=testing_cls_integers,
                          y_pred=cls_predicted)

    # Printing the confusion matrix
    for i in range(num_classes):
        # Append the class-name to each line.
        class_name = "({}) {}".format(i, class_names[i])
        print(cm[i, :], class_name)

    # labeling each column of the confusion matrix with the class number
    cls_numbers = [" ({0})".format(i) for i in range(num_classes)]
```

```
    print("".join(cls_numbers))
```

Also, we are going to define another helper function to run the trained classifier over the test set and measure the accuracy of the trained model over the test set:

```
# Split the data-set in batches of this size to limit RAM usage.
batch_size = 128

def predict_class(transferValues, labels, cls_true):
    # Number of images.
    num_imgs = len(transferValues)

    # Allocate an array for the predicted classes which
    # will be calculated in batches and filled into this array.
    cls_predicted = np.zeros(shape=num_imgs, dtype=np.int)

    # Now calculate the predicted classes for the batches.
    # We will just iterate through all the batches.
    # There might be a more clever and Pythonic way of doing this.

    # The starting index for the next batch is denoted i.
    i = 0

    while i < num_imgs:
        # The ending index for the next batch is denoted j.
        j = min(i + batch_size, num_imgs)

        # Create a feed-dict with the images and labels
        # between index i and j.
        feed_dict = {input_values: transferValues[i:j],
                     y_actual: labels[i:j]}

        # Calculate the predicted class using TensorFlow.
        cls_predicted[i:j] = session.run(y_predicted_cls,
feed_dict=feed_dict)

        # Set the start-index for the next batch to the
        # end-index of the current batch.
        i = j
    # Create a boolean array whether each image is correctly classified.
    correct = [a == p for a, p in zip(cls_true, cls_predicted)]

    return correct, cls_predicted

#Calling the above function making the predictions for the test

def predict_cls_test():
    return predict_class(transferValues = transfer_values_test,
```

```
                        labels = labels_test,
                        cls_true = cls_test)

def classification_accuracy(correct):
    # When averaging a boolean array, False means 0 and True means 1.
    # So we are calculating: number of True / len(correct) which is
    # the same as the classification accuracy.

    # Return the classification accuracy
    # and the number of correct classifications.
    return np.mean(correct), np.sum(correct)

def test_accuracy(show_example_errors=False,
                  show_confusion_matrix=False):

    # For all the images in the test-set,
    # calculate the predicted classes and whether they are correct.
    correct, cls_pred = predict_class_test()
    # Classification accuracypredict_class_test and the number of correct
classifications.
    accuracy, num_correct = classification_accuracy(correct)
    # Number of images being classified.
    num_images = len(correct)

    # Print the accuracy.
    msg = "Test set accuracy: {0:.1%} ({1} / {2})"
    print(msg.format(accuracy, num_correct, num_images))

    # Plot some examples of mis-classifications, if desired.
    if show_example_errors:
        print("Example errors:")
        plot_errors(cls_predicted=cls_pred, cls_correct=correct)

    # Plot the confusion matrix, if desired.
    if show_confusion_matrix:
        print("Confusion Matrix:")
        plot_confusionMatrix(cls_predicted=cls_pred)
```

Let's see the performance of the previous neural network model before doing any optimization:

```
test_accuracy(show_example_errors=True,
              show_confusion_matrix=True)

Accuracy on Test-Set: 9.4% (939 / 10000)
```

As you can see, the performance of the network is very low, but it will get better after doing some optimization based on the optimization criteria that we already defined. So we are going to run the optimizer for 10,000 iterations and test the model accuracy after that:

```
optimize(num_iterations=10000)
test_accuracy(show_example_errors=True,
                    show_confusion_matrix=True)
Accuracy on Test-Set: 90.7% (9069 / 10000)
Example errors:
```

Figure 10.11: Some misclassified images from the test set

```
Confusion Matrix:
[926    6   13    2    3    0    1    1   29   19]  (0) airplane
[   9  921    2    5    0    1    1    1    2   58]  (1) automobile
[  18    1  883   31   32    4   22    5    1    3]  (2) bird
[   7    2   19  855   23   57   24    9    2    2]  (3) cat
[   5    0   21   25  896    4   24   22    2    1]  (4) deer
[   2    0   12   97   18  843   10   15    1    2]  (5) dog
[   2    1   16   17   17    4  940    1    2    0]  (6) frog
[   8    0   10   19   28   14    1  914    2    4]  (7) horse
[  42    6    1    4    1    0    2    0  932   12]  (8) ship
[   6   19    2    2    1    0    1    1    9  959]  (9) truck
  (0)  (1)  (2)  (3)  (4)  (5)  (6)  (7)  (8)  (9)
```

To wrap this up, we are going to close the opened sessions:

```
model.close()
session.close()
```

Summary

In this chapter, we introduced one of the most widely used best practices of deep learning. TL is a very exciting tool that you can use to get deep learning architectures to learn from your small dataset, but make sure you use it in the right way.

Next up, we are going to introduce a widely used deep learning architecture for natural language processing. These recurrent-type architectures have achieved a breakthrough in most NLP domains: machine translation, speech recognition, language modeling, and sentiment analysis.

10
Recurrent-Type Neural Networks - Language Modeling

Recurrent neural networks (**RNNs**) are a class of deep learning architectures that are widely used for natural language processing. This set of architectures enables us to provide contextual information for current predictions and also have specific architecture that deals with long-term dependencies in any input sequence. In this chapter, we'll demonstrate how to make a sequence-to-sequence model, which will be useful in many applications in NLP. We will demonstrate these concepts by building a character-level language model and see how our model generates sentences similar to original input sequences.

The following topics will be covered in this chapter:

- The intuition behind RNNs
- LSTM networks
- Implementation of the language model

The intuition behind RNNs

All the deep learning architectures that we have dealt with so far have no mechanism to memorize the input that they have received previously. For instance, if you feed a **feed-forward neural network** (**FNN**) with a sequence of characters such as **HELLO**, when the network gets to **E**, you will find that it didn't preserve any information/forgotten that it just read **H**. This is a serious problem for sequence-based learning. And since it has no memory of any previous characters it read, this kind of network will be very difficult to train to predict the next character. This doesn't make sense for lots of applications such as language modeling, machine translation, speech recognition, and so on.

For this specific reason, we are going to introduce RNNs, a set of deep learning architectures that do preserve information and memorize what they have just encountered.

Let's demonstrate how RNNs should work on the same input sequence of characters, **HELLO**. When the RNN cell/unit receives **E** as an input, it also receives that character **H**, which it received earlier. This feeding of the present character along with the past one as an input to the RNN cell gives a great advantage to these architectures, which is short-term memory; it also makes these architectures usable for predicting/guessing the most likely character after **H**, which is **L**, in this specific sequence of characters.

We have seen that previous architectures assign weights to their inputs; RNNs follow the same optimization process of assigning weights to their multiple inputs, which is the present and past. So in this case, the network will assign two different matrices of weights to each one of them. In order to do that, we will be using **gradient descent** and a heavier version of backpropagation, which is called **backpropagation through time (BPTT)**.

Recurrent neural networks architectures

Depending on our background of using previous deep learning architectures, you will find out why RNNs are special. The previous architectures that we have learned about are not flexible in terms of their input or training. They accept a fixed-size sequence/vector/image as an input and produce another fixed-size one as an output. RNN architectures are somehow different, because they enable you to feed a sequence as input and get another sequence as output, or to have sequences in the input only/output only as shown in *Figure 1*. This kind of flexibility is very useful for multiple applications such as language modeling and sentiment analysis:

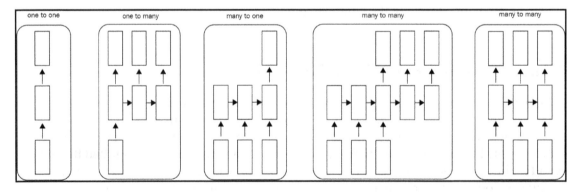

Figure 1: Flexibility of RNNs in terms of shape of input or output (http://karpathy.github.io/2015/05/21/rnn-effectiveness/)

The intuition behind these set of architectures is to mimic the way humans process information. In any typical conversation your understanding of someone's words is totally dependent on what he said previously and you might even be able to predict what he's going to say next based on what he just said.

The exact same process should be followed in the case of RNNs. For example, imagine you want translate a specific word in a sentence. You can't use traditional FNNs for that, because they won't be able to use the translation of previous words as an input with the current word that we want to translate, and this may result in an incorrect translation because of the lack of contextual information around this word.

RNNs do preserves information about the past and they have some kind of loops to allow the previously learned information to be used for the current prediction at any given point:

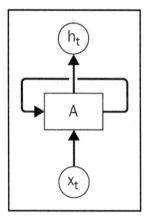

Figure 2: RNNs architecture which has loop to persist information for past steps (source: http://colah.github.io/posts/2015-08-Understanding-LSTMs/)

In *Figure 2*, we have some neural networks called *A* which receives an input X_t and produces and output h_t. Also, it receives information from past steps with the help of this loop.

This loop seems to unclear, but if we used the unrolled version of *Figure 2*, you will find out that it's very simple and intuitive, and that the RNN is nothing but a repeated version of the same network (which could be normal FNN), as shown in *Figure 3*:

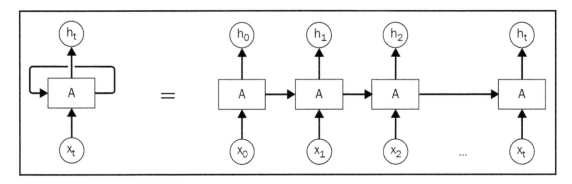

Figure 3: An unrolled version of the recurrent neural network architecture (source: http://colah.github.io/posts/2015-08-Understanding-LSTMs/)

This intuitive architecture of RNNs and its flexibility in terms of input/output shape make them a good fit for interesting sequence-based learning tasks such as machine translation, language modeling, sentiment analysis, image captioning, and more.

Examples of RNNs

Now, we have an intuitive understanding of how RNNs work and how it's going to be useful in different interesting sequence-based examples. Let's have a closer look of some of these interesting examples.

Character-level language models

Language modeling is an essential task for many applications such as speech recognition, machine translation and more. In this section, we'll try to mimic the training process of RNNs and get a deeper understanding of how these networks work. We'll build a language model that operate over characters. So, we will feed our network with a chunk of text with the purpose of trying to build a probability distribution of the next character given the previous ones which will allow us to generate text similar to the one we feed as an input in the training process.

For example, suppose we have a language with only four letters as its vocabulary, **helo**.

The task is to train a recurrent neural network on a specific input sequence of characters such as **hello**. In this specific example, we have four training samples:

1. The probability of the character **e** should be calculated given the context of the first input character **h**,
2. The probability of the character l should be calculated given the context of **he**,
3. The probability of the character l should be calculated given the context of **hel**, and
4. Finally the probability of the character **o** should be calculated given the context of **hell**

As we learned in previous chapters, machine learning techniques in general which deep learning is a part of, only accept real-value numbers as input. So, wee need somehow convert or encode or input character to a numerical form. To do this, we will use one-hot-vector encoding which is a way to encode text by have a vector of zeros except for a single entry in the vector, which is the index of the character in the vocabulary of this language that we are trying to model (in this case **helo**). After encoding our training samples, we will provide them to the RNN-type model one at a time. At each given character, the output of the RNN-type model will be a 4-dimensional vector (the size of the vector corresponds to the size of the vocab) which represents the probability of each character in the vocabulary being the next one after the given input character. *Figure 4* clarifies this process:

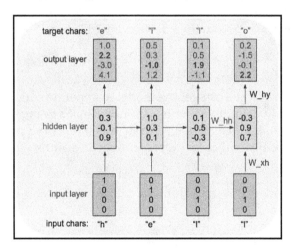

Figure 4: Example of RNN-type network with one-hot-vector encoded characters as an input and the output will be distribution over the vocab representing the most likely character after the current one (source: http://karpathy.github.io/2015/05/21/rnn-effectiveness/)

As shown in *Figure 4*, you can see that we fed the first character in our input sequence **h** to the model and the output was 4-dimensional vector representing the confidence about the next character. So it has a confidence of **1.0** of **h** being the next character after the input **h**, a confidence of **2.2** of **e** being the next character, a confidence of **-3.0** to **l** being the next character, and finally a confidence of **4.1** to **o** being the next character. In this specific example, we know the correct next character will be **e**, based on our training sequence **hello**. So our primary goal while training this RNN-type network is increase the confidence of **e** being the next character and decrease the confidence of other characters. To do this kind of optimization we will be using gradient descent and backpropagation algorithms to update the weights and influence the network to produce a higher confidence for our correct next character, **e**, and so on, for the other 3 training examples.

As you can see the output of the RNN-type network produces a confidence distribution over all the characters of the vocab being the next one. We can turn this confidence distribution into a probability distribution such that the increase of one characters probability being the next one will result in decreasing the others probabilities because the probability needs to sum up to 1. For this specific modification we can use a standard softmax layer to every output vector.

For generating text from these kind of networks, we can feed an initial character to the model and get a probability distribution over the characters that are likely to be next, and then we can sample from these characters and feed it back as an input to the model. We'll be able to get a sequence of characters by repeating this process over and over again as many times as we want to generate a text with a desired length.

Language model using Shakespeare data

From the preceding example, we can get the model to generate text. But the network will surprise us, as it's not only going to generate text but also it's going to learn the style and the structure in training data. We can demonstrate this interesting process by training an RNN-type model on specific kind of text that has structure and style in it, such as the following Shakespeare work.

Let's have a look at a generated output from the trained network:

> *Second Senator:*
> *They are away this miseries, produced upon my soul,*
> *Breaking and strongly should be buried, when I perish*
> *The earth and thoughts of many states.*

In spite of the fact that the network only knows how to produce one single character at a time, it was able to generate a meaningful text and names that actually have the structure and style of Shakespeare work.

The vanishing gradient problem

While training these sets of RNN-type architectures, we use gradient descent and backprogagation through time, which introduced some successes for lots of sequence-based learning tasks. But because of the nature of the gradient and due to using fast training strategies, it could be shown that the gradient values will tend to be too small and vanish. This process introduced the vanishing gradient problem that many practitioners fall into. Later on in this chapter, we will discuss how researchers approached these kind of problems and produced variations of the vanilla RNNs to overcome this problem:

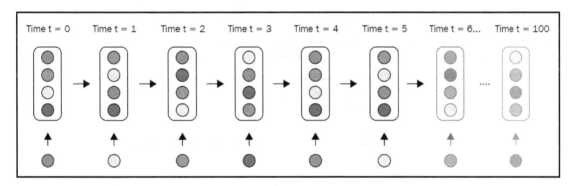

Figure 5: Vanishing gradient problem

The problem of long-term dependencies

Another challenging problem faced by researchers is the long-term dependencies that one can find in text. For example, if someone feeds a sequence like *I used to live in France and I learned how to speak...* the next obvious word in the sequence is the word French.

In these kind of situation vanilla RNNs will be able to handle it because it has short-term dependencies, as shown in *Figure 6*:

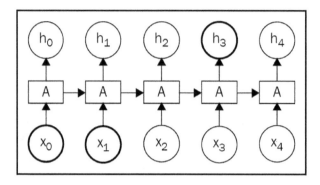

Figure 6: Showing short-term dependencies in the text (source: http://colah.github.io/posts/2015-08-Understanding-LSTMs/)

Another example, if someone started the sequence by saying that *I used to live in France..* and then he/she start to describe the beauty of living there and finally he ended the sequence by *I learned to speak French*. So, for the model to predict the language that he/she learned at the end of the sequence, the model needs to have some information about the early words *live* and *France*. The model won't be able to handle these kind of situation, if it doesn't manage to keep track of long term dependencies in the text:

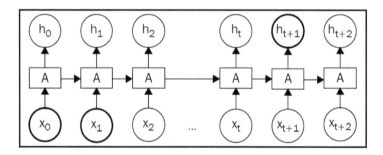

Figure 7: The challenge of long-term dependencies in text (source: http://colah.github.io/posts/2015-08-Understanding-LSTMs/)

To handle vanishing gradients and long-term dependencies in the text, researchers introduced a variation of the vanilla RNN network called **Long Short Term Networks (LSTM)**.

LSTM networks

LSTM, a variation of an RNN that is used to help learning long term dependencies in the text. LSTMs were initially introduced by Hochreiter & Schmidhuber (1997) (link: `http://www.bioinf.jku.at/publications/older/2604.pdf`), and many researchers worked on it and produced interesting results in many domains.

These kind of architectures will be able to handle the problem of long-term dependencies in the text because of its inner architecture.

LSTMs are similar to the vanilla RNN as it has a repeating module over time, but the inner architecture of this repeated module is different from the vanilla RNNs. It includes more layers for forgetting and updating information:

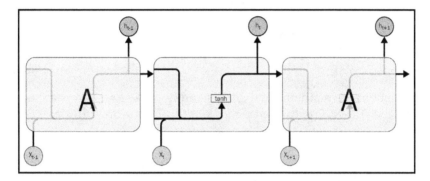

Figure 8: The repeating module in a standard RNN containing a single layer (source: http://colah.github.io/posts/2015-08-Understanding-LSTMs/)

As mentioned previously, the vanilla RNNs have a single NN layer, but the LSTMs have four different layers interacting in a special way. This special kind of interaction is what makes LSTM, work very well for many domains, which we'll see while building our language model example:

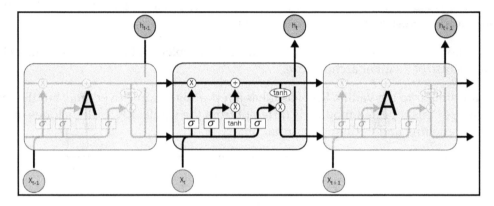

Figure 9: The repeating module in an LSTM containing four interacting layers (source: http://colah.github.io/posts/2015-08-Understanding-LSTMs/)

For more details about the mathematical details and how the four layers are actually interacting with each other, you can have a look at this interesting tutorial: `http://colah.github.io/posts/2015-08-Understanding-LSTMs/`

Why does LSTM work?

The first step in our vanilla LSTM architecture it to decide which information is not necessary and it will work by throwing it away to leave more room for more important information. For this, we have a layer called **forget gate layer**, which looks at the previous output h_{t-1} and the current input x_t and decides which information we are going to throw away.

The next step in the LSTM architecture is to decide which information is worth keeping/persisting and storing in the cell. This is done in two steps:

1. A layer called **input gate layer**, which decides which values of the previous state of the cell needs to be updated
2. The second step is to generate a set of new candidate values that will be added to the cell

Finally, we need to decide what the LSTM cell is going to output. This output will be based on our cell state, but will be a filtered version.

Implementation of the language model

In this section, we'll build a language model that operates over characters. For this implementation, we will use an Anna Karenina novel and see how the network will learn to implement the structure and style of the text:

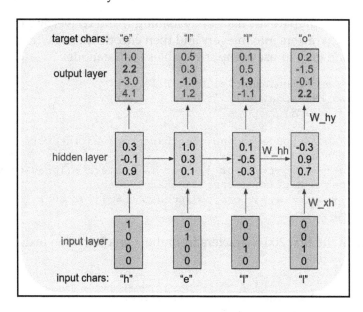

Figure 10: General architecture for the character-level RNN (source: http://karpathy.github.io/2015/05/21/rnn-effectiveness/)

This network is based off of Andrej Karpathy's post on RNNs (link: `http://karpathy.github.io/2015/05/21/rnn-effectiveness/`) and implementation in Torch (link: `https://github.com/karpathy/char-rnn`). Also, there's some information here at r2rt (link: `http://r2rt.com/recurrent-neural-networks-in-tensorflow-ii.html`) and from Sherjil Ozairp (link: `https://github.com/sherjilozair/char-rnn-tensorflow`) on GitHub. The following is the general architecture of the character-wise RNN.

We'll build a character-level RNN trained on the Anna Karenina novel (link: `https://en.wikipedia.org/wiki/Anna_Karenina`). It'll be able to generate new text based on the text from the book. You will find the `.txt` file included with the assets of this implementation.

Let's start by importing the necessary libraries for this character-level implementation:

```
import numpy as np
import tensorflow as tf

from collections import namedtuple
```

To start off, we need to prepare the dataset by loading it and converting it in to integers. So, we will convert the characters into integers and then encode them as integers which makes it straightforward and easy to use as input variables for the model:

```
#reading the Anna Karenina novel text file
with open('Anna_Karenina.txt', 'r') as f:
    textlines=f.read()

#Building the vocan and encoding the characters as integers
language_vocab = set(textlines)
vocab_to_integer = {char: j for j, char in enumerate(language_vocab)}
integer_to_vocab = dict(enumerate(language_vocab))
encoded_vocab = np.array([vocab_to_integer[char] for char in textlines],
dtype=np.int32)
```

So, let's have look at the first 200 characters from the Anna Karenina text:

```
textlines[:200]
Output:
"Chapter 1\n\n\nHappy families are all alike; every unhappy family is
unhappy in its own\nway.\n\nEverything was in confusion in the Oblonskys'
house. The wife had\ndiscovered that the husband was carrying on"
```

We have also converted the characters to a convenient form for the network, which is integers. So, let's have a look at the encoded version of the characters:

```
encoded_vocab[:200]
Output:
array([70, 34, 54, 29, 24, 19, 76, 45,  2, 79, 79, 79, 69, 54, 29, 29, 49,
       45, 66, 54, 39, 15, 44, 15, 19, 12, 45, 54, 76, 19, 45, 54, 44, 44,
       45, 54, 44, 15, 27, 19, 58, 45, 19, 30, 19, 76, 49, 45, 59, 56, 34,
       54, 29, 29, 49, 45, 66, 54, 39, 15, 44, 49, 45, 15, 12, 45, 59, 56,
       34, 54, 29, 29, 49, 45, 15, 56, 45, 15, 24, 12, 45, 11, 35, 56, 79,
       35, 54, 49, 53, 79, 79, 36, 30, 19, 76, 49, 24, 34, 15, 56, 16, 45,
       35, 54, 12, 45, 15, 56, 45, 31, 11, 56, 66, 59, 12, 15, 11, 56, 45,
       15, 56, 45, 24, 34, 19, 45,  1, 82, 44, 11, 56, 12, 27, 49, 12, 37,
       45, 34, 11, 59, 12, 19, 53, 45, 21, 34, 19, 45, 35, 15, 66, 19, 45,
       34, 54, 64, 79, 64, 15, 12, 31, 11, 30, 19, 76, 19, 64, 45, 24, 34,
       54, 24, 45, 24, 34, 19, 45, 34, 59, 12, 82, 54, 56, 64, 45, 35, 54,
       12, 45, 31, 54, 76, 76, 49, 15, 56, 16, 45, 11, 56], dtype=int32)
```

Since the network is working with individual characters, it's similar to a classification problem in which we are trying to predict the next character from the previous text. Here's how many classes our network has to pick from.

So, we will be feeding the model a character at a time, and the model will predict the next character by producing a probability distribution over the possible number of characters that could come next (vocab), which is equivalent to a number of classes the network needs to pick from:

```
len(language_vocab)
Output:
83
```

Since we'll be using stochastic gradient descent to train our model, we need to convert our data into training batches.

Mini-batch generation for training

In this section, we will divide our data into small batches to be used for training. So, the batches will consist of many sequences of desired number of sequence steps. So, let's look at a visual example in *Figure 11*:

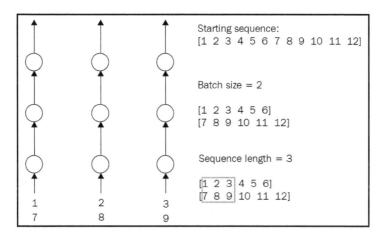

Figure 11: Illustration of how batches and sequences would look like (source: http://oscarmore2.github.io/Anna_KaRNNa_files/charseq.jpeg)

So, now we need to define a function that will iterate through the encoded text and generate the batches. In this function we will be using a very nice mechanism of Python called **yield** (link: https://jeffknupp.com/blog/2013/04/07/improve-your-python-yield-and-generators-explained/).

A typical batch will have $N \times M$ characters, where N is the number of sequences and M is, number of sequence steps. For getting the number of possible batches in our dataset, we can simply divide the length of the data by the desired batch size and after getting this number of possible batches, we can drive how many characters should be in each batch.

After that, we need to split the dataset we have into a desired number of sequences (N). We can use `arr.reshape(size)`. We know we want N sequences (`num_seqs` is used in, following code), let's make that the size of the first dimension. For the second dimension, you can use -1 as a placeholder in the size; it'll fill up the array with the appropriate data for you. After this, you should have an array that is $N \times (M * K)$, where K is the number of batches.

Now that we have this array, we can iterate through it to get the training batches, where each batch has $N \times M$ characters. For each subsequent batch, the window moves over by `num_steps`. Finally, we also want to create both the input and output arrays for ours to be used as the model input. This step of creating the output values is very easy; remember that the targets are the inputs shifted over one character. You'll usually see the first input character used as the last target character, so something like this:

$$y[:, : -1], y[:, -1] = x[:, 1 :], x[:, 0]$$

Where x is the input batch and y is the target batch.

The way I like to do this window is to use range to take steps of size `num_steps`, starting from 0 to `arr.shape[1]`, the total number of steps in each sequence. That way, the integers you get from the range always point to the start of a batch, and each window is `num_steps` wide:

```
def generate_character_batches(data, num_seq, num_steps):
    '''Create a function that returns batches of size
       num_seq x num_steps from data.
    '''
    # Get the number of characters per batch and number of batches
    num_char_per_batch = num_seq * num_steps
    num_batches = len(data)//num_char_per_batch
    # Keep only enough characters to make full batches
    data = data[:num_batches * num_char_per_batch]
    # Reshape the array into n_seqs rows
    data = data.reshape((num_seq, -1))
    for i in range(0, data.shape[1], num_steps):
        # The input variables
        input_x = data[:, i:i+num_steps]
        # The output variables which are shifted by one
```

```
            output_y = np.zeros_like(input_x)
            output_y[:, :-1], output_y[:, -1] = input_x[:, 1:], input_x[:, 0]
            yield input_x, output_y
```

So, let's demonstrate this using this function by generating a batch of 15 sequences and 50 sequence steps:

```
generated_batches = generate_character_batches(encoded_vocab, 15, 50)
input_x, output_y = next(generated_batches)
print('input\n', input_x[:10, :10])
print('\ntarget\n', output_y[:10, :10])
Output:

input
 [[70 34 54 29 24 19 76 45 2 79]
 [45 19 44 15 16 15 82 44 19 45]
 [11 45 44 15 16 34 24 38 34 19]
 [45 34 54 64 45 82 19 19 56 45]
 [45 11 56 19 45 27 56 19 35 79]
 [49 19 54 76 12 45 44 54 12 24]
 [45 41 19 45 16 11 45 15 56 24]
 [11 35 45 24 11 45 39 54 27 19]
 [82 19 66 11 76 19 45 81 19 56]
 [12 54 16 19 45 44 15 27 19 45]]

target
 [[34 54 29 24 19 76 45 2 79 79]
 [19 44 15 16 15 82 44 19 45 16]
 [45 44 15 16 34 24 38 34 19 54]
 [34 54 64 45 82 19 19 56 45 82]
 [11 56 19 45 27 56 19 35 79 35]
 [19 54 76 12 45 44 54 12 24 45]
 [41 19 45 16 11 45 15 56 24 11]
 [35 45 24 11 45 39 54 27 19 33]
 [19 66 11 76 19 45 81 19 56 24]
 [54 16 19 45 44 15 27 19 45 24]]
```

Next up, we'll be looking forward to building the core of this example, which is the LSTM model.

Building the model

Before diving into building the character-level model using LSTMs, it is worth mentioning something called **Stacked LSTM**.

Stacked LSTMs are useful for looking at your information at different time scales.

Stacked LSTMs

> *"Building a deep RNN by stacking multiple recurrent hidden states on top of each other. This approach potentially allows the hidden state at each level to operate at different timescale" How to Construct Deep Recurrent Neural Networks (link: https://arxiv.org/abs/1312.6026), 2013*

> *"RNNs are inherently deep in time, since their hidden state is a function of all previous hidden states. The question that inspired this paper was whether RNNs could also benefit from depth in space; that is from stacking multiple recurrent hidden layers on top of each other, just as feedforward layers are stacked in conventional deep networks". Speech Recognition With Deep RNNs (link:* `https://arxiv.org/abs/1303.5778`*), 2013*

Most researchers are using stacked LSTMs for challenging sequence prediction problems. A stacked LSTM architecture can be defined as an LSTM model comprised of multiple LSTM layers. The preceding LSTM layer provides a sequence output rather than a single-value output to the LSTM layer as follows.

Specifically, it's one output per input time step, rather than one output time step for all input time steps:

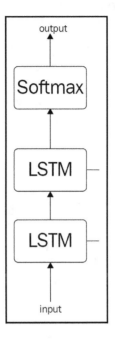

Figure 12: Stacked LSTMs

So in this example, we will be using this kind of stacked LSTM architecture, which gives better performance.

Model architecture

This is where you'll build the network. We'll break it into parts so that it's easier to reason about each bit. Then, we can connect them with the whole network:

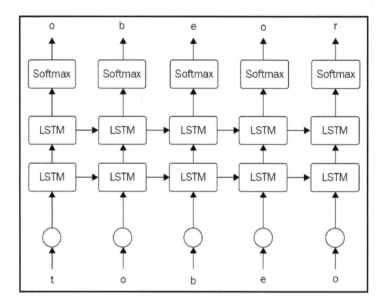

Figure 13: Character-level model architecture

Inputs

Now, let's start by defining the model inputs as placeholders. The inputs of the model will be training data and the targets. We will also use a parameter called `keep_probability` for the dropout layer, which helps the model avoid overfitting:

```
def build_model_inputs(batch_size, num_steps):
    # Declare placeholders for the input and output variables
    inputs_x = tf.placeholder(tf.int32, [batch_size, num_steps],
name='inputs')
    targets_y = tf.placeholder(tf.int32, [batch_size, num_steps],
name='targets')
    # define the keep_probability for the dropout layer
    keep_probability = tf.placeholder(tf.float32, name='keep_prob')
    return inputs_x, targets_y, keep_probability
```

Building an LSTM cell

In this section, we will write a function for creating the LSTM cell, which will be used in the hidden layer. This cell will be the building block for our model. So, we will create this cell using TensorFlow. Let's have a look at how we can use TensorFlow to build a basic LSTM cell.

We call the following line of code to create an LSTM cell with the parameter num_units representing the number of units in the hidden layer:

```
lstm_cell = tf.contrib.rnn.BasicLSTMCell(num_units)
```

To prevent overfitting, we can use something called **dropout**, which is a mechanism for preventing the model from overfitting the data by decreasing the model's complexity:

```
tf.contrib.rnn.DropoutWrapper(lstm, output_keep_prob=keep_probability)
```

As we mentioned before, we will be using the stacked LSTM architecture; it will help us to look at the data from different angles and has been practically found to perform better. In order to define a stacked LSTM in TensorFlow, we can use the tf.contrib.rnn.MultiRNNCell function (link: https://www.tensorflow.org/versions/r1.0/api_docs/python/tf/contrib/rnn/MultiRNNCell):

```
tf.contrib.rnn.MultiRNNCell([cell]*num_layers)
```

Initially for the first cell, there will be no previous information, so we need to initialize the cell state to be zeros. We can use the following function to do that:

```
initial_state = cell.zero_state(batch_size, tf.float32)
```

So, let's put it all together and create our LSTM cell:

```
def build_lstm_cell(size, num_layers, batch_size, keep_probability):

    ### Building the LSTM Cell using the tensorflow function
    lstm_cell = tf.contrib.rnn.BasicLSTMCell(size)
    # Adding dropout to the layer to prevent overfitting
    drop_layer = tf.contrib.rnn.DropoutWrapper(lstm_cell,
output_keep_prob=keep_probability)
    # Add muliple cells together and stack them up to oprovide a level of
more understanding
    stakced_cell = tf.contrib.rnn.MultiRNNCell([drop_layer] * num_layers)
    initial_cell_state = lstm_cell.zero_state(batch_size, tf.float32)
    return lstm_cell, initial_cell_state
```

RNN output

Next up, we need to create the output layer, which is responsible for reading the output of the individual LSTM cells and passing them through a fully connected layer. This layer has a softmax output for producing a probability distribution over the likely character to be next after the input one.

As you know, we have generated input batches for the network with size $N \times M$ characters, where N is the number of sequences in this batch and M is the number of sequence steps. We have also used L hidden units in the hidden layer while creating the model. Based on the batch size and number of hidden units, the output of the network will be a *3D* Tensor with size $N \times M \times L$, and that's because we call the LSTM cell M times, one for each sequence step. Each call to LSTM cell produces an output of size L. Finally, we need to do this as many as number of sequences N as the we have.

So we pass this $N \times M \times L$ output to a fully connected layer (which is the same for all outputs with the same weights), but before doing this, we reshape the output to a *2D* tensor, which has a shape of *(M * N) × L*. This reshaping will make things easier for us when operating on the output, because the new shape will be more convenient; the values of each row represents the L outputs of the LSTM cell, and hence it's one row for each sequence and step.

After getting the new shape, we can connect it to the fully connected layer with the softmax by doing matrix multiplication with the weights. The weights created in the LSTM cells and the weight that we are about to create here have the same name by default, and TensorFlow will raise an error in such a case. To avoid this error, we can wrap the weight and bias variables created here in a variable scope using the TensorFlow function `tf.variable_scope()`.

After explaining the shape of the output and how we are going to reshape it, to make things easier, let's go ahead and code this `build_model_output` function:

```
def build_model_output(output, input_size, output_size):

    # Reshaping output of the model to become a bunch of rows, where each
row correspond for each step in the seq
    sequence_output = tf.concat(output, axis=1)
    reshaped_output = tf.reshape(sequence_output, [-1, input_size])
    # Connect the RNN outputs to a softmax layer
    with tf.variable_scope('softmax'):
        softmax_w = tf.Variable(tf.truncated_normal((input_size,
output_size), stddev=0.1))
        softmax_b = tf.Variable(tf.zeros(output_size))
    # the output is a set of rows of LSTM cell outputs, so the logits will
```

```
be a set
    # of rows of logit outputs, one for each step and sequence
    logits = tf.matmul(reshaped_output, softmax_w) + softmax_b
    # Use softmax to get the probabilities for predicted characters
    model_out = tf.nn.softmax(logits, name='predictions')
    return model_out, logits
```

Training loss

Next up is the training loss. We get the logits and targets and calculate the softmax cross-entropy loss. First, we need to one-hot encode the targets; we're getting them as encoded characters. Then, we reshape the one-hot targets, so it's a *2D* tensor with size $(M * N) \times C$, where C is the number of classes/characters we have. Remember that we reshaped the LSTM outputs and ran them through a fully connected layer with C units. So, our logits will also have size $(M * N) \times C$.

Then, we run the `logits` and `targets` through
`tf.nn.softmax_cross_entropy_with_logits` and find the mean to get the loss:

```
def model_loss(logits, targets, lstm_size, num_classes):
    # convert the targets to one-hot encoded and reshape them to match the
logits, one row per batch_size per step
    output_y_one_hot = tf.one_hot(targets, num_classes)
    output_y_reshaped = tf.reshape(output_y_one_hot, logits.get_shape())
    #Use the cross entropy loss
    model_loss = tf.nn.softmax_cross_entropy_with_logits(logits=logits,
labels=output_y_reshaped)
    model_loss = tf.reduce_mean(model_loss)
    return model_loss
```

Optimizer

Finally, we need to use an optimization method that will help us learn something from the dataset. As we know, vanilla RNNs have exploding and vanishing gradient issues. LSTMs fix only one issue, which is the vanishing of the gradient values, but even after using LSTM, some gradient values explode and grow without bounds. In order to fix this problem, we can use something called **gradient clipping**, which is a technique to clip the gradients that explode to a specific threshold.

So, let's define our optimizer by using the Adam optimizer for the learning process:

```
def build_model_optimizer(model_loss, learning_rate, grad_clip):
```

```
    # define optimizer for training, using gradient clipping to avoid the
exploding of the gradients
    trainable_variables = tf.trainable_variables()
    gradients, _ = tf.clip_by_global_norm(tf.gradients(model_loss,
trainable_variables), grad_clip)
    #Use Adam Optimizer
    train_operation = tf.train.AdamOptimizer(learning_rate)
    model_optimizer = train_operation.apply_gradients(zip(gradients,
trainable_variables))
    return model_optimizer
```

Building the network

Now, we can put all the pieces together and build a class for the network. To actually run data through the LSTM cells, we will use `tf.nn.dynamic_rnn` (link: `https://www.tensorflow.org/versions/r1.0/api_docs/python/tf/nn/dynamic_rnn`). This function will pass the hidden and cell states across LSTM cells appropriately for us. It returns the outputs for each LSTM cell at each step for each sequence in the mini-batch. It also gives us the final LSTM state. We want to save this state as `final_state`, so we can pass it to the first LSTM cell in the the next mini-batch run. For `tf.nn.dynamic_rnn`, we pass in the cell and initial state we get from `build_lstm`, as well as our input sequences. Also, we need to one-hot encode the inputs before going into the RNN:

```
class CharLSTM:
    def __init__(self, num_classes, batch_size=64, num_steps=50,
                       lstm_size=128, num_layers=2, learning_rate=0.001,
                       grad_clip=5, sampling=False):
        # When we're using this network for generating text by sampling,
we'll be providing the network with
        # one character at a time, so providing an option for it.
        if sampling == True:
            batch_size, num_steps = 1, 1
        else:
            batch_size, num_steps = batch_size, num_steps

        tf.reset_default_graph()
        # Build the model inputs placeholders of the input and target
variables
        self.inputs, self.targets, self.keep_prob =
build_model_inputs(batch_size, num_steps)

        # Building the LSTM cell
        lstm_cell, self.initial_state = build_lstm_cell(lstm_size,
num_layers, batch_size, self.keep_prob)
```

```
### Run the data through the LSTM layers
# one_hot encode the input
input_x_one_hot = tf.one_hot(self.inputs, num_classes)
# Runing each sequence step through the LSTM architecture and
finally collecting the outputs
outputs, state = tf.nn.dynamic_rnn(lstm_cell, input_x_one_hot,
initial_state=self.initial_state)
self.final_state = state
# Get softmax predictions and logits
self.prediction, self.logits = build_model_output(outputs,
lstm_size, num_classes)
# Loss and optimizer (with gradient clipping)
self.loss = model_loss(self.logits, self.targets, lstm_size,
num_classes)
self.optimizer = build_model_optimizer(self.loss, learning_rate,
grad_clip)
```

Model hyperparameters

As with any deep learning architecture, there are a few hyperparameters that someone can use to control the model and fine-tune it. The following is the set of hyperparameters that we are using for this architecture:

- Batch size is the number of sequences running through the network in one pass.
- The number of steps is the number of characters in the sequence the network is trained on. Larger is better typically; the network will learn more long-range dependencies, but will take longer to train. 100 is typically a good number here.
- The LSTM size is the number of units in the hidden layers.
- Architecture number layers is the number of hidden LSTM layers to use.
- Learning rate is the typical learning rate for training.
- And finally, the new thing that we call keep probability is used by the dropout layer; it helps the network to avoid overfitting. So if your network is overfitting, try decreasing this.

Training the model

Now, let's kick off the training process by providing the inputs and outputs to the built model and then use the optimizer to train the network. Don't forget that we need to use the previous state while making predictions for the current state. Thus, we need to pass the output state back to the network so that it can be used during the prediction of the next input.

Let's provide initial values for our hyperparameters (you can tune them afterwards depending on the dataset you are using to train this architecture):

```
batch_size = 100        # Sequences per batch
num_steps = 100         # Number of sequence steps per batch
lstm_size = 512         # Size of hidden layers in LSTMs
num_layers = 2          # Number of LSTM layers
learning_rate = 0.001   # Learning rate
keep_probability = 0.5  # Dropout keep probability

epochs = 5

# Save a checkpoint N iterations
save_every_n = 100

LSTM_model = CharLSTM(len(language_vocab), batch_size=batch_size,
num_steps=num_steps,
                lstm_size=lstm_size, num_layers=num_layers,
                learning_rate=learning_rate)

saver = tf.train.Saver(max_to_keep=100)
with tf.Session() as sess:
    sess.run(tf.global_variables_initializer())
    # Use the line below to load a checkpoint and resume training
    #saver.restore(sess, 'checkpoints/_____.ckpt')
    counter = 0
    for e in range(epochs):
        # Train network
        new_state = sess.run(LSTM_model.initial_state)
        loss = 0
        for x, y in generate_character_batches(encoded_vocab, batch_size,
num_steps):
            counter += 1
            start = time.time()
            feed = {LSTM_model.inputs: x,
                    LSTM_model.targets: y,
                    LSTM_model.keep_prob: keep_probability,
                    LSTM_model.initial_state: new_state}
            batch_loss, new_state, _ = sess.run([LSTM_model.loss,
                                                LSTM_model.final_state,
                                                LSTM_model.optimizer],
                                                feed_dict=feed)
            end = time.time()
            print('Epoch number: {}/{}... '.format(e+1, epochs),
                  'Step: {}... '.format(counter),
                  'loss: {:.4f}... '.format(batch_loss),
                  '{:.3f} sec/batch'.format((end-start)))
            if (counter % save_every_n == 0):
```

```
            saver.save(sess, "checkpoints/i{}_l{}.ckpt".format(counter,
lstm_size))
        saver.save(sess, "checkpoints/i{}_l{}.ckpt".format(counter, lstm_size))
```

At the end of the training process, you should get an error close to this:

```
    .
    .
    .
Epoch number: 5/5... Step: 978... loss: 1.7151... 0.050 sec/batch
Epoch number: 5/5... Step: 979... loss: 1.7428... 0.051 sec/batch
Epoch number: 5/5... Step: 980... loss: 1.7151... 0.050 sec/batch
Epoch number: 5/5... Step: 981... loss: 1.7236... 0.050 sec/batch
Epoch number: 5/5... Step: 982... loss: 1.7314... 0.051 sec/batch
Epoch number: 5/5... Step: 983... loss: 1.7369... 0.051 sec/batch
Epoch number: 5/5... Step: 984... loss: 1.7075... 0.065 sec/batch
Epoch number: 5/5... Step: 985... loss: 1.7304... 0.051 sec/batch
Epoch number: 5/5... Step: 986... loss: 1.7128... 0.049 sec/batch
Epoch number: 5/5... Step: 987... loss: 1.7107... 0.051 sec/batch
Epoch number: 5/5... Step: 988... loss: 1.7351... 0.051 sec/batch
Epoch number: 5/5... Step: 989... loss: 1.7260... 0.049 sec/batch
Epoch number: 5/5... Step: 990... loss: 1.7144... 0.051 sec/batch
```

Saving checkpoints

Now, let's load the checkpoints. For more about saving and loading checkpoints, you can check out the TensorFlow documentation (https://www.tensorflow.org/programmers_guide/variables):

```
tf.train.get_checkpoint_state('checkpoints')

Output:
model_checkpoint_path: "checkpoints/i990_l512.ckpt"
all_model_checkpoint_paths: "checkpoints/i100_l512.ckpt"
all_model_checkpoint_paths: "checkpoints/i200_l512.ckpt"
all_model_checkpoint_paths: "checkpoints/i300_l512.ckpt"
all_model_checkpoint_paths: "checkpoints/i400_l512.ckpt"
all_model_checkpoint_paths: "checkpoints/i500_l512.ckpt"
all_model_checkpoint_paths: "checkpoints/i600_l512.ckpt"
all_model_checkpoint_paths: "checkpoints/i700_l512.ckpt"
all_model_checkpoint_paths: "checkpoints/i800_l512.ckpt"
all_model_checkpoint_paths: "checkpoints/i900_l512.ckpt"
all_model_checkpoint_paths: "checkpoints/i990_l512.ckpt"
```

Generating text

We have a trained model based on our input dataset. The next step is to use this trained model to generate text and see how this model learned the style and structure of the input data. To do this, we can start with some initial characters and then feed the new, predicted one as an input in the next step. We will repeat this process until we get a text with a specific length.

In the following code, we have also added extra statements to the function to prime the network with some initial text and start from there.

The network gives us predictions or probabilities for each character in the vocab. To reduce noise and only use the ones that the network is more confident about, we're going to only choose a new character from the top *N* most probable characters in the output:

```python
def choose_top_n_characters(preds, vocab_size, top_n_chars=4):
    p = np.squeeze(preds)
    p[np.argsort(p)[:-top_n_chars]] = 0
    p = p / np.sum(p)
    c = np.random.choice(vocab_size, 1, p=p)[0]
    return c

def sample_from_LSTM_output(checkpoint, n_samples, lstm_size, vocab_size,
prime="The "):
    samples = [char for char in prime]
    LSTM_model = CharLSTM(len(language_vocab), lstm_size=lstm_size,
sampling=True)
    saver = tf.train.Saver()
    with tf.Session() as sess:
        saver.restore(sess, checkpoint)
        new_state = sess.run(LSTM_model.initial_state)
        for char in prime:
            x = np.zeros((1, 1))
            x[0,0] = vocab_to_integer[char]
            feed = {LSTM_model.inputs: x,
                    LSTM_model.keep_prob: 1.,
                    LSTM_model.initial_state: new_state}
            preds, new_state = sess.run([LSTM_model.prediction,
LSTM_model.final_state],
                                        feed_dict=feed)

        c = choose_top_n_characters(preds, len(language_vocab))
        samples.append(integer_to_vocab[c])

        for i in range(n_samples):
            x[0,0] = c
```

```
feed = {LSTM_model.inputs: x,
        LSTM_model.keep_prob: 1.,
        LSTM_model.initial_state: new_state}
preds, new_state = sess.run([LSTM_model.prediction,
LSTM_model.final_state],
                                    feed_dict=feed)

c = choose_top_n_characters(preds, len(language_vocab))
samples.append(integer_to_vocab[c])
return ''.join(samples)
```

Let's start the sampling process using the latest checkpoint saved:

```
tf.train.latest_checkpoint('checkpoints')

Output:
'checkpoints/i990_l512.ckpt'
```

Now, it's time to sample using this latest checkpoint:

```
checkpoint = tf.train.latest_checkpoint('checkpoints')
sampled_text = sample_from_LSTM_output(checkpoint, 1000, lstm_size,
len(language_vocab), prime="Far")
print(sampled_text)

Output:
INFO:tensorflow:Restoring parameters from checkpoints/i990_l512.ckpt

Farcial the
confiring to the mone of the correm and thinds. She
she saw the
streads of herself hand only astended of the carres to her his some of the
princess of which he came him of
all that his white the dreasing of
thisking the princess and with she was she had
bettee a still and he was happined, with the pood on the mush to the
peaters and seet it.

"The possess a streatich, the may were notine at his mate a misted
and the
man of the mother at the same of the seem her
felt. He had not here.

"I conest only be alw you thinking that the partion
of their said."

"A much then you make all her
somether. Hower their centing
```

```
about
this, and I won't give it in
himself.
I had not come at any see it will that there she chile no one that him.

"The distiction with you all.... It was
a mone of the mind were starding to the simple to a mone. It to be to ser
in the place," said Vronsky.
"And a plais in
his face, has alled in the consess on at they to gan in the sint
at as that
he would not be and t
```

You can see that we were able to generate some meaningful words and some meaningless words. In order to get more results, you can run the model for more epochs and try to play with the hyperparameters.

Summary

We learned about RNNs, how they work, and why they have become a big deal. We trained an RNN character-level language model on fun novel datasets and saw where RNNs are going. You can confidently expect a large amount of innovation in the space of RNNs, and I believe they will become a pervasive and critical component of intelligent systems.

11
Representation Learning - Implementing Word Embeddings

Machine learning is a science that is mainly based on statistics and linear algebra. Applying matrix operations is very common among most machine learning or deep learning architectures because of backpropagation. This is the main reason deep learning, or machine learning in general, accepts only real-valued quantities as input. This fact contradicts many applications, such as machine translation, sentiment analysis, and so on; they have text as an input. So, in order to use deep learning for this application, we need to have it in the form that deep learning accepts!

In this chapter, we are going to introduce the field of representation learning, which is a way to learn a real-valued representation from text while preserving the semantics of the actual text. For example, the representation of love should be very close to the representation of adore because they are used in very similar contexts.

So, the following topics will be covered in this chapter:

- Introduction to representation learning
- Word2Vec
- A practical example of the skip-gram architecture
- Skip-gram Word2Vec implementation

Introduction to representation learning

All the machine learning algorithms or architectures that we have used so far require the input to be real-valued or matrices of real-valued quantities, and that's a common theme in machine learning. For example, in the convolution neural network, we had to feed raw pixel values of images as model inputs. In this part, we are dealing with text, so we need to encode our text somehow and produce real-valued quantities that can be fed to a machine learning algorithm. In order to encode input text as real-valued quantities, we need to use an intermediate science called **Natural Language Processing (NLP)**.

We mentioned that in this kind of pipeline, where we feed text to a machine learning model such as sentiment analysis, this will be problematic and won't work because we won't be able to apply backpropagation or any other operations such as dot product on the input, which is a string. So, we need to use a mechanism of NLP that will enable us to build an intermediate representation of the text that can carry the same information as the text and also be fed to the machine learning models.

We need to convert each word or token in the input text to a real-valued vector. These vectors will be useless if they don't carry the patterns, information, meaning, and semantics of the original input. For example, as in real text, the two words love and adore are very similar to each other and carry the same meaning. We need the resultant real-valued vectors that will represent them to be close to each other and be in the same vector space. So, the vector representation of these two words along with another word that isn't similar to them will be like this diagram:

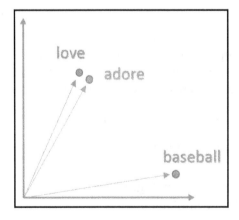

Figure 15.1: Vector representation of words

There are many techniques that can be used for this task. This family of techniques is called **embeddings**, where you're embedding text into another real-valued vector space.

As we'll see later on, this vector space is very interesting actually, because you will find out that you can drive a word's vectors from other words that are similar to it or even do some geography in this space.

Word2Vec

Word2Vec is one of the widely used embedding techniques in the area of NLP. This model creates real-valued vectors from input text by looking at the contextual information the input word appears in. So, you will find out that similar words will be mentioned in very similar contexts, and hence the model will learn that those two words should be placed close to each other in the particular embedding space.

From the statements in the following diagram, the model will learn that the words **love** and **adore** share very similar contexts and should be placed very close to each other in the resulting vector space. The context of like could be a bit similar as well to the word love, but it won't be as close to love as the word adore:

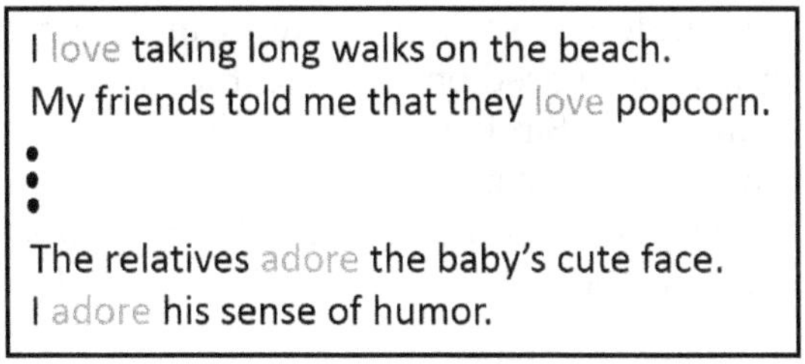

I love **taking long walks on the beach.**
My friends told me that they love **popcorn.**

⋮

The relatives adore **the baby's cute face.**
I adore **his sense of humor.**

Figure 15.2: Sample of sentiment sentences

The Word2Vec model also relies on semantic features of input sentences; for example, the two words adore and love are mainly used in a positive context and usually precede noun phrases or nouns. Again, the model will learn that these two words have something in common and it will be more likely to put the vector representation of these two vectors in a similar context. So, the structure of the sentence will tell the Word2Vec model a lot about similar words.

In practice, people feed a large corpus of text to the Word2Vec model. The model will learn to produce similar vectors for similar words, and it will do so for each unique word in the input text.

All of these words' vectors will be combined and the final output will be an embedding matrix where each row represents the real-valued vector representation of a specific unique word.

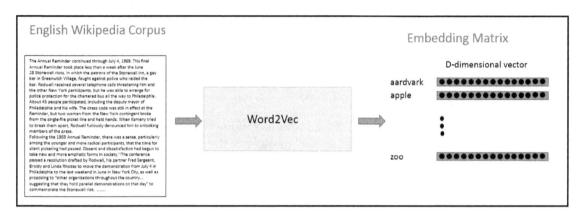

Figure 15.3: Example of Word2Vec model pipeline

So, the final output of the model will be an embedding matrix for all the unique words in the training corpus. Usually, good embedding matrices could contain millions of real-valued vectors.

Word2Vec modeling uses a window to scan the sentence and then tries to predict the vector of the middle word of that window based on its contextual information; the Word2Vec model will scan a sentence at a time. Similar to any machine learning technique, we need to define a cost function for the Word2Vec model and its corresponding optimization criteria that will make the model capable of generating real-valued vectors for each unique image and also relate the vectors to each other based on their contextual information

Building Word2Vec model

In this section, we will go through some deeper details of how can we build a Word2Vec model. As we mentioned previously, our final goal is to have a trained model that will able to generate real-valued vector representation for the input textual data which is also called word embeddings.

During the training of the model, we will use the maximum likelihood method (https://en.wikipedia.org/wiki/Maximum_likelihood), which can be used to maximize the probability of the next word w_t in the input sentence given the previous words that the model has seen, which we can call h.

This maximum likelihood method will be expressed in terms of the softmax function:

$$P(w_t|h) = \text{softmax}(\text{score}(w_t, h))$$
$$= \frac{\exp\{\text{score}(w_t, h)\}}{\sum_{\text{Word w' in Vocab}} \exp\{\text{score}(w', h)\}}$$

Here, the *score* function computes a value to represent the compatibility of the target word w_t with respect to the context h. This model will be trained on the input sequences while training to maximize the likelihood on the training input data (log likelihood is used for mathematical simplicity and derivation with the log):

$$J_{\text{ML}} = \log P(w_t|h)$$
$$= \text{score}(w_t, h) - \log\left(\sum_{\text{Word w' in Vocab}} \exp\{\text{score}(w', h)\}\right).$$

So, the *ML* method will try to maximize the above equation which, will result in a probabilistic language model. But the calculation of this is very computationally expensive, as we need to compute each probability using the score function for all the words in the vocabulary *V* words w', in the corresponding current context h of this model. This will happen at every training step.

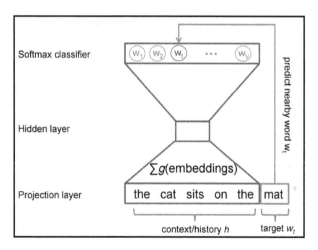

Figure 15.4: General architecture of a probabilistic language model

Because of the computational expensiveness of building the probabilistic language model, people tend to use different techniques that are less computationally expensive, such as **Continuous Bag-of-Words (CBOW)** and skip-gram models.

These models are trained to build a binary classification with logistic regression to separate between the real target words w_t and h noise or imaginary words $\tilde{w}$, which is in the same context. The following diagram simplifies this idea using the CBOW technique:

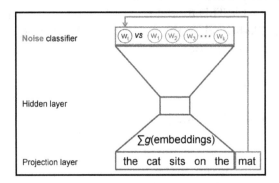

Figure 15.5: General architecture of skip-gram model

The next diagram, shows the two architectures that you can use for building the Word2Vec model:

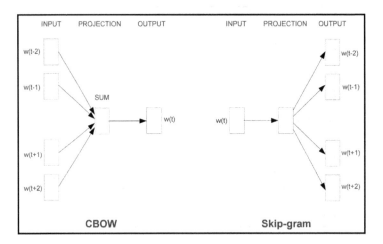

Figure 15.6: different architectures for the Word2Vec model

To be more formal, the objective function of these techniques maximizes the following:

$$J_{\text{NEG}} = \log Q_\theta(D = 1|w_t, h) + k \underset{\tilde{w} \sim P_{\text{noise}}}{\mathbb{E}} \left[\log Q_\theta(D = 0|\tilde{w}, h)\right]$$

Where:

- $Q_\theta(D = 1|w, h)$ is the probability of the binary logistic regression based on the model seeing the word w in the context h in the dataset D, which is calculated in terms of the θ vector. This vector represents the learned embeddings.

- $\tilde{w}$ is the imaginary or noisy words that we can generate from a noisy probabilistic distribution, such as the unigram of the training input examples.

To sum up, the objective of these models is to discriminate between real and imaginary inputs, and hence assign higher probability to real words and less probability for the case of imaginary or noisy words.

This objective is maximized when the model assigns high probabilities to real words and low probabilities to noise words.

Technically, the process of assigning high probability to real words is is called **negative sampling** (https://papers.nips.cc/paper/5021-distributed-representations-of-words-and-phrases-and-their-compositionality.pdf), and there is good mathematical motivation for using this loss function: the updates it proposes approximate the updates of the softmax function in the limit. But computationally, it is especially appealing because computing the loss function now scales only with the number of noise words that we select (k), and not all words in the vocabulary (V). This makes it much faster to train. We will actually make use of the very similar **noise-contrastive estimation** (**NCE**) (https://papers.nips.cc/paper/5165-learning-word-embeddings-efficiently-with-noise-contrastive-estimation.pdf) loss, for which TensorFlow has a handy helper function, `tf.nn.nce_loss()`.

A practical example of the skip-gram architecture

Let's go through a practical example and see how skip-gram models will work in this situation:

```
the quick brown fox jumped over the lazy dog
```

First off, we need to make a dataset of words and their corresponding context. Defining the context is up to us, but it has to make sense. So, we'll take a window around the target word and take a word from the right and another from the left.

By following this contextual technique, we will end up with the following set of words and their corresponding context:

```
([the, brown], quick), ([quick, fox], brown), ([brown, jumped], fox), ...
```

The generated words and their corresponding context will be represented as pairs of `(context, target)`. The idea of skip-gram models is the inverse of CBOW ones. In the skip- gram model, we will try to predict the context of the word based on its target word. For example, considering the first pair, the skip-gram model will try to predict `the` and `brown` from the target word `quick`, and so on. So, we can rewrite our dataset as follows:

```
(quick, the), (quick, brown), (brown, quick), (brown, fox), ...
```

Now, we have a set of input and output pairs.

Let's try to mimic the training process at specific step t. So, the skip-gram model will take the first training sample where the input is the word `quick` and the target output is the word `the`. Next, we need to construct the noisy input as well, so we are going to randomly select from the unigrams of the input data. For simplicity, the size of the noisy vector will be only one. For example, we can select the word `sheep` as a noisy example.

Now, we can go ahead and compute the loss between the real pair and the noisy one as:

$$J_{\mathrm{NEG}}^{(t)} = \log Q_\theta(D = 1|\text{the, quick}) + \log(Q_\theta(D = 0|\text{sheep, quick}))$$

The goal in this case is to update the θ parameter to improve the previous objective function. Typically, we can use gradient for this. So, we will try to calculate the gradient of the loss with respect to the objective function parameter θ, which will be represented by $\frac{\partial}{\partial \theta} J_{\text{NEG}}$.

After the training process, we can visualize some results based on their reduced dimensions of the real-valued vector representation. You will find that this vector space is very interesting because you can do lots of interesting stuff with it. For example, you can learn Analogy in this space by saying that king is to queen as man is to woman. We can even derive the woman vector by subtracting the king vector from the queen one and adding the man; the result of this will be very close to the actual learned vector of the woman. You can also learn geography in this space.

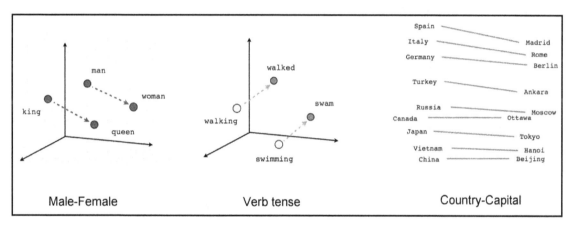

Figure 15.7: Projection of the learned vectors to two dimensions using t-distributed stochastic neighbor embedding (t-SNE) dimensionality reduction technique

The preceding example gives very good intuition behind these vectors and how they'll be useful for most NLP applications such as machine translation or **part-of-speech** (**POS**) tagging.

Skip-gram Word2Vec implementation

After understanding the mathematical details of how skip-gram models work, we are going to implement skip-gram, which encodes words into real-valued vectors that have certain properties (hence the name Word2Vec). By implementing this architecture, you will get a clue of how the process of learning another representation works.

Text is the main input for a lot of natural language processing applications such as machine translation, sentiment analysis, and text to speech systems. So, learning a real-valued representation for the text will help us use different deep learning techniques for these tasks.

In the early chapters of this book, we introduced something called one-hot encoding, which produces a vector of zeros except for the index of the word that this vector represents. So, you may wonder why we are not using it here. This method is very inefficient because usually you have a big set of distinct words, maybe something like 50,000 words, and using one-hot encoding for this will produce a vector of 49,999 entries set to zero and only one entry set to one.

Having a very sparse input like this will result in a huge waste of computation because of the matrix multiplications that we'd do in the hidden layers of the neural network.

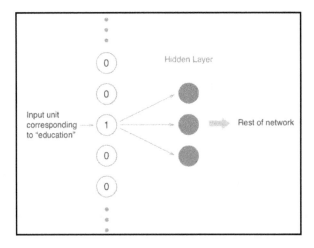

Figure 15.8: One-hot encoding which will result in huge waste of computation

As we mentioned previously, the outcome of using one-hot encoding will be a very sparse vector, especially when you have a huge amount of distinct words that you want to encode.

The following figure shows that when we multiply this sparse vector of all zeros except for one entry by a matrix of weights, the output will be only the row of the matrix that corresponds to the one value of the sparse vector:

$$\begin{bmatrix} 0 & 0 & 0 & 1 & 0 \end{bmatrix} \times \begin{bmatrix} 8 & 2 & 1 & 9 \\ 6 & 5 & 4 & 0 \\ 7 & 1 & 6 & 2 \\ 1 & 3 & 5 & 8 \\ 0 & 4 & 9 & 1 \end{bmatrix} = \begin{bmatrix} 1 & 3 & 5 & 8 \end{bmatrix}$$

Figure 15.9: The effect of multiplying a one-hot vector with almost all zeros by hidden layer weight matrix

To avoid this huge waste of computation, we will be using embeddings, which is just a fully-connected layer with some embedding weights. In this layer, we skip this inefficient multiplication and look up the embedding weights of the embedding layer from something called **weight matrix**.

So, instead of the waste that results from the computation, we are going to use this weight lookup this weight matrix to find the embedding weights. First, need to build this lookup take. To do this, we are going to encode all the input words as integers, as shown in the following figure, and then to get the corresponding values for this word, we are going to use its integer representation as the row number in this weight matrix. The process of finding the corresponding embedding values of a specific word is called **embedding lookup.** As mentioned previously, the embedding layer will be just a fully connected layer, where the number of units represents the embedding dimension.

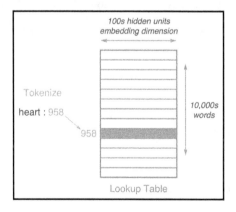

Figure 15.10: Tokenized lookup table

You can see that this process is very intuitive and straightforward; we just need to follow these steps:

1. Define the lookup table that will be considered as a weight matrix
2. Define the embedding layer as a fully connected hidden layer with specific number of units (embedding dimensions)
3. Use the weight matrix lookup as an alternative for the computationally unnecessary matrix multiplication
4. Finally, train the lookup table as any weight matrix

As we mentioned earlier, we are going to build a skip-gram Word2Vec model in this section, which is an efficient way of learning a representation for words while preserving the semantic information that the words have.

So, let's go ahead and build a Word2Vec model using the skip-gram architecture, which is proven to better than others.

Data analysis and pre-processing

In this section, we are going to define some helper functions that will enable us to build a good Word2Vec model. For this implementation, we are going to use a cleaned version of Wikipedia (http://mattmahoney.net/dc/textdata.html).

So, let's start off by importing the required packages for this implementation:

```
#importing the required packages for this implementation
import numpy as np
import tensorflow as tf


#Packages for downloading the dataset
from urllib.request import urlretrieve
from os.path import isfile, isdir
from tqdm import tqdm
import zipfile

#packages for data preprocessing
import re
from collections import Counter
import random
```

Next up, we are going to define a class that will be used to download the dataset if it was not downloaded before:

```
# In this implementation we will use a cleaned up version of Wikipedia from
Matt Mahoney.
# So we will define a helper class that will helps to download the dataset
wiki_dataset_folder_path = 'wikipedia_data'
wiki_dataset_filename = 'text8.zip'
wiki_dataset_name = 'Text8 Dataset'

class DLProgress(tqdm):
    last_block = 0

    def hook(self, block_num=1, block_size=1, total_size=None):
        self.total = total_size
        self.update((block_num - self.last_block) * block_size)
        self.last_block = block_num
# Cheking if the file is not already downloaded
if not isfile(wiki_dataset_filename):
    with DLProgress(unit='B', unit_scale=True, miniters=1,
desc=wiki_dataset_name) as pbar:
        urlretrieve(
            'http://mattmahoney.net/dc/text8.zip',
            wiki_dataset_filename,
            pbar.hook)

# Checking if the data is already extracted if not extract it
if not isdir(wiki_dataset_folder_path):
    with zipfile.ZipFile(wiki_dataset_filename) as zip_ref:
        zip_ref.extractall(wiki_dataset_folder_path)
with open('wikipedia_data/text8') as f:
    cleaned_wikipedia_text = f.read()

Output:

Text8 Dataset: 31.4MB [00:39, 794kB/s]
```

We can have a look at the first 100 characters of this dataset:

```
cleaned_wikipedia_text[0:100]

' anarchism originated as a term of abuse first used against early working
class radicals including t'
```

Next up, we are going to preprocess the text, so we are going to define a helper function that will help us to replace special characters such as punctuation ones into a know token. Also, to reduce the amount of noise in the input text, you might want to remove words that don't appear frequently in the text:

```
def preprocess_text(input_text):

    # Replace punctuation with some special tokens so we can use them in
our model
    input_text = input_text.lower()
    input_text = input_text.replace('.', ' <PERIOD> ')
    input_text = input_text.replace(',', ' <COMMA> ')
    input_text = input_text.replace('"', ' <QUOTATION_MARK> ')
    input_text = input_text.replace(';', ' <SEMICOLON> ')
    input_text = input_text.replace('!', ' <EXCLAMATION_MARK> ')
    input_text = input_text.replace('?', ' <QUESTION_MARK> ')
    input_text = input_text.replace('(', ' <LEFT_PAREN> ')
    input_text = input_text.replace(')', ' <RIGHT_PAREN> ')
    input_text = input_text.replace('--', ' <HYPHENS> ')
    input_text = input_text.replace('?', ' <QUESTION_MARK> ')
    input_text = input_text.replace(':', ' <COLON> ')
    text_words = input_text.split()
    # neglecting all the words that have five occurrences of fewer
    text_word_counts = Counter(text_words)
    trimmed_words = [word for word in text_words if text_word_counts[word]
> 5]

    return trimmed_words
```

Now, let's call this function on the input text and have a look at the output:

```
preprocessed_words = preprocess_text(cleaned_wikipedia_text)
print(preprocessed_words[:30])

Output:
['anarchism', 'originated', 'as', 'a', 'term', 'of', 'abuse', 'first',
'used', 'against', 'early', 'working', 'class', 'radicals', 'including',
'the', 'diggers', 'of', 'the', 'english', 'revolution', 'and', 'the',
'sans', 'culottes', 'of', 'the', 'french', 'revolution', 'whilst']
```

Let's see how many words and distinct words we have for the pre-processed version of the text:

```
print("Total number of words in the text:
{}".format(len(preprocessed_words)))
print("Total number of unique words in the text:
{}".format(len(set(preprocessed_words))))
```

```
Output:

Total number of words in the text: 16680599
Total number of unique words in the text: 63641
```

And here, I'm creating dictionaries to covert words to integers and backwards, that is, integers to words. The integers are assigned in descending frequency order, so the most frequent word (the) is given the integer 0, the next most frequent gets 1, and so on. The words are converted to integers and stored in the list int_words.

As mentioned earlier in this section, we need to use the integer indexes of the words to look up their values in the weight matrix, so we are going to words to integers and integers to words. This will help us to look up the words and also get the actual word of a specific index. For example, the most repeated word in the input text will be indexed at position 0, followed by the second most repeated one, and so on.

So, let's define a function to create this lookup table:

```python
def create_lookuptables(input_words):
    """
    Creating lookup tables for vocan

    Function arguments:
    param words: Input list of words
    """
    input_word_counts = Counter(input_words)
    sorted_vocab = sorted(input_word_counts, key=input_word_counts.get,
reverse=True)
    integer_to_vocab = {ii: word for ii, word in enumerate(sorted_vocab)}
    vocab_to_integer = {word: ii for ii, word in integer_to_vocab.items()}

    # returning A tuple of dicts
    return vocab_to_integer, integer_to_vocab
```

Now, let's call the defined function to create the lookup table:

```python
vocab_to_integer, integer_to_vocab =
create_lookuptables(preprocessed_words)
integer_words = [vocab_to_integer[word] for word in preprocessed_words]
```

To build a more accurate model, we can remove words that don't change the context much as of, for, the, and so on. So, it is practically proven that we can build more accurate models while discarding these kinds of words. The process of removing context-irrelevant words from the context is called **subsampling**. In order to define a general mechanism for word discarding, Mikolov introduced a function for calculating the discard probability of a certain word, which is given by:

$$P(w_i) = 1 - \sqrt{\frac{t}{f(w_i)}}$$

Where:

- t is a threshold parameter for word discarding
- $f(w_i)$ is the frequency of a specific target word w_i in the input dataset

So, we are going to implement a helper function that will calculate the discarding probability of each word in the dataset:

```
# removing context-irrelevant words threshold
word_threshold = 1e-5

word_counts = Counter(integer_words)
total_number_words = len(integer_words)

#Calculating the freqs for the words
frequencies = {word: count/total_number_words for word, count in
word_counts.items()}

#Calculating the discard probability
prob_drop = {word: 1 - np.sqrt(word_threshold/frequencies[word]) for word
in word_counts}
training_words = [word for word in integer_words if random.random() < (1 -
prob_drop[word])]
```

Now, we have a more refined and clean version of the input text.

We mentioned that the skip-gram architecture considers the context of the target word while producing its real-valued representation, so it defines a window around the target word that has size C.

Instead of treating all contextual words equally, we are going to assign less weight for words that are a bit far from the target word. For example, if we choose the size of the window to be $C = 4$, then we are going to select a random number L from the range of 1 to C, and then sample L words from the history and the future of the current word. For more details about this, refer to the Mikolov et al paper at: https://arxiv.org/pdf/1301.3781.pdf.

So, let's go ahead and define this function:

```
# Defining a function that returns the words around specific index in a
specific window
def get_target(input_words, ind, context_window_size=5):
    #selecting random number to be used for genearting words form history
and feature of the current word
    rnd_num = np.random.randint(1, context_window_size+1)
    start_ind = ind - rnd_num if (ind - rnd_num) > 0 else 0
    stop_ind = ind + rnd_num
    target_words = set(input_words[start_ind:ind] +
input_words[ind+1:stop_ind+1])
    return list(target_words)
```

Also, let's define a generator function to generate a random batch from the training samples and get the contextual word for each word in that batch:

```
#Defining a function for generating word batches as a tuple (inputs,
targets)
def generate_random_batches(input_words, train_batch_size,
context_window_size=5):
    num_batches = len(input_words)//train_batch_size
    # working on only only full batches
    input_words = input_words[:num_batches*train_batch_size]
    for ind in range(0, len(input_words), train_batch_size):
        input_vals, target = [], []
        input_batch = input_words[ind:ind+train_batch_size]
        #Getting the context for each word
        for ii in range(len(input_batch)):
            batch_input_vals = input_batch[ii]
            batch_target = get_target(input_batch, ii, context_window_size)
            target.extend(batch_target)
            input_vals.extend([batch_input_vals]*len(batch_target))
        yield input_vals, target
```

Building the model

Next up, we are going to use the following structure to build the computational graph:

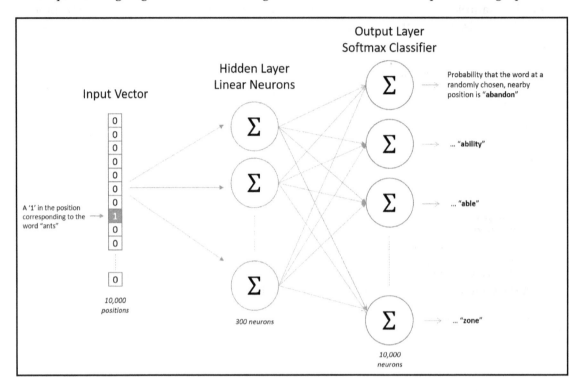

Figure 15.11: Model architecture

So, as mentioned previously, we are going to use an embedding layer that will try to learn a special real-valued representation for these words. Thus, the words will be fed as one-hot vectors. The idea is to train this network to build up the weight matrix.

So, let's start off by creating the input to our model:

```
train_graph = tf.Graph()

#defining the inputs placeholders of the model
with train_graph.as_default():
    inputs_values = tf.placeholder(tf.int32, [None], name='inputs_values')
    labels_values = tf.placeholder(tf.int32, [None, None],
name='labels_values')
```

The weight or embedding matrix that we are trying to build will have the following shape:

```
num_words X num_hidden_neurons
```

Also, we don't have to implement the lookup function ourselves because it's already available in Tensorflow: `tf.nn.embedding_lookup()`. So, it will use the integer encoding of the words and locate their corresponding rows in the weight matrix.

The weight matrix will be randomly initialized from a uniform distribution:

```
num_vocab = len(integer_to_vocab)

num_embedding =  300
with train_graph.as_default():
    embedding_layer = tf.Variable(tf.random_uniform((num_vocab,
num_embedding), -1, 1))
    # Next, we are going to use tf.nn.embedding_lookup function to get the
output of the hidden layer
    embed_tensors = tf.nn.embedding_lookup(embedding_layer, inputs_values)
```

It's very inefficient to update all the embedding weights of the embedding layer at once. Instead of this, we will use the negative sampling technique which will only update the weight of the correct word with a small subset of the incorrect ones.

Also, we don't have to implement this function ourselves as it's already there in TensorFlow `tf.nn.sampled_softmax_loss`:

```
# Number of negative labels to sample
num_sampled = 100

with train_graph.as_default():
    # create softmax weights and biases
    softmax_weights = tf.Variable(tf.truncated_normal((num_vocab,
num_embedding)))
    softmax_biases = tf.Variable(tf.zeros(num_vocab), name="softmax_bias")
    # Calculating the model loss using negative sampling
    model_loss = tf.nn.sampled_softmax_loss(
        weights=softmax_weights,
        biases=softmax_biases,
        labels=labels_values,
        inputs=embed_tensors,
        num_sampled=num_sampled,
        num_classes=num_vocab)
    model_cost = tf.reduce_mean(model_loss)
    model_optimizer = tf.train.AdamOptimizer().minimize(model_cost)
```

To validate our trained model, we are going to sample some frequent or common words and some uncommon words and try to print our their closest set of words based on the learned representation of the skip-gram architecture:

```
with train_graph.as_default():
    # set of random words for evaluating similarity on
    valid_num_words = 16
    valid_window = 100
    # pick 8 samples from (0,100) and (1000,1100) each ranges. lower id
implies more frequent
    valid_samples = np.array(random.sample(range(valid_window),
valid_num_words//2))
    valid_samples = np.append(valid_samples,
                             random.sample(range(1000,1000+valid_window),
valid_num_words//2))
    valid_dataset_samples = tf.constant(valid_samples, dtype=tf.int32)
    # Calculating the cosine distance
    norm = tf.sqrt(tf.reduce_sum(tf.square(embedding_layer), 1,
keep_dims=True))
    normalized_embed = embedding_layer / norm
    valid_embedding = tf.nn.embedding_lookup(normalized_embed,
valid_dataset_samples)
    cosine_similarity = tf.matmul(valid_embedding,
tf.transpose(normalized_embed))
```

Now, we have all the bits and pieces for our model and we are ready to kick off the training process.

Training

Let's go ahead and kick off the training process:

```
num_epochs = 10
train_batch_size = 1000
contextual_window_size = 10

with train_graph.as_default():
    saver = tf.train.Saver()

with tf.Session(graph=train_graph) as sess:
    iteration_num = 1
    average_loss = 0
    #Initializing all the vairables
    sess.run(tf.global_variables_initializer())
```

```
    for e in range(1, num_epochs+1):
        #Generating random batch for training
        batches = generate_random_batches(training_words, train_batch_size,
contextual_window_size)
        #Iterating through the batch samples
        for input_vals, target in batches:
            #Creating the feed dict
            feed_dict = {inputs_values: input_vals,
                    labels_values: np.array(target)[:, None]}
            train_loss, _ = sess.run([model_cost, model_optimizer],
feed_dict=feed_dict)
            #commulating the loss
            average_loss += train_loss
            #Printing out the results after 100 iteration
            if iteration_num % 100 == 0:
                print("Epoch Number {}/{}".format(e, num_epochs),
                    "Iteration Number: {}".format(iteration_num),
                    "Avg. Training loss:
{:.4f}".format(average_loss/100))
                average_loss = 0
            if iteration_num % 1000 == 0:
                ## Using cosine similarity to get the nearest words to a
word
                similarity = cosine_similarity.eval()
                for i in range(valid_num_words):
                    valid_word = integer_to_vocab[valid_samples[i]]
                    # number of nearest neighbors
                    top_k = 8
                    nearest_words = (-similarity[i,
:]).argsort()[1:top_k+1]
                    msg = 'The nearest to %s:' % valid_word
                    for k in range(top_k):
                        similar_word = integer_to_vocab[nearest_words[k]]
                        msg = '%s %s,' % (msg, similar_word)
                    print(msg)
            iteration_num += 1
    save_path = saver.save(sess,
"checkpoints/cleaned_wikipedia_version.ckpt")
    embed_mat = sess.run(normalized_embed)
```

After running the preceding code snippet for 10 epochs, you will get the following output:

```
Epoch Number 10/10 Iteration Number: 43100 Avg. Training loss: 5.0380
Epoch Number 10/10 Iteration Number: 43200 Avg. Training loss: 4.9619
Epoch Number 10/10 Iteration Number: 43300 Avg. Training loss: 4.9463
Epoch Number 10/10 Iteration Number: 43400 Avg. Training loss: 4.9728
Epoch Number 10/10 Iteration Number: 43500 Avg. Training loss: 4.9872
Epoch Number 10/10 Iteration Number: 43600 Avg. Training loss: 5.0534
```

```
Epoch Number 10/10 Iteration Number: 43700 Avg. Training loss: 4.8261
Epoch Number 10/10 Iteration Number: 43800 Avg. Training loss: 4.8752
Epoch Number 10/10 Iteration Number: 43900 Avg. Training loss: 4.9818
Epoch Number 10/10 Iteration Number: 44000 Avg. Training loss: 4.9251
The nearest to nine: one, seven, zero, two, three, four, eight, five,
The nearest to such: is, as, or, some, have, be, that, physical,
The nearest to who: his, him, he, did, to, had, was, whom,
The nearest to two: zero, one, three, seven, four, five, six, nine,
The nearest to which: as, a, the, in, to, also, for, is,
The nearest to seven: eight, one, three, five, four, six, zero, two,
The nearest to american: actor, nine, singer, actress, musician, comedian,
athlete, songwriter,
The nearest to many: as, other, some, have, also, these, are, or,
The nearest to powers: constitution, constitutional, formally, assembly,
state, legislative, general, government,
The nearest to question: questions, existence, whether, answer, truth,
reality, notion, does,
The nearest to channel: tv, television, broadcasts, broadcasting, radio,
channels, broadcast, stations,
The nearest to recorded: band, rock, studio, songs, album, song, recording,
pop,
The nearest to arts: art, school, alumni, schools, students, university,
renowned, education,
The nearest to orthodox: churches, orthodoxy, church, catholic, catholics,
oriental, christianity, christians,
The nearest to scale: scales, parts, important, note, between, its, see,
measured,
The nearest to mean: is, exactly, defined, denote, hence, are, meaning,
example,

Epoch Number 10/10 Iteration Number: 45100 Avg. Training loss: 4.8466
Epoch Number 10/10 Iteration Number: 45200 Avg. Training loss: 4.8836
Epoch Number 10/10 Iteration Number: 45300 Avg. Training loss: 4.9016
Epoch Number 10/10 Iteration Number: 45400 Avg. Training loss: 5.0218
Epoch Number 10/10 Iteration Number: 45500 Avg. Training loss: 5.1409
Epoch Number 10/10 Iteration Number: 45600 Avg. Training loss: 4.7864
Epoch Number 10/10 Iteration Number: 45700 Avg. Training loss: 4.9312
Epoch Number 10/10 Iteration Number: 45800 Avg. Training loss: 4.9097
Epoch Number 10/10 Iteration Number: 45900 Avg. Training loss: 4.6924
Epoch Number 10/10 Iteration Number: 46000 Avg. Training loss: 4.8999
The nearest to nine: one, eight, seven, six, four, five, american, two,
The nearest to such: can, example, examples, some, be, which, this, or,
The nearest to who: him, his, himself, he, was, whom, men, said,
The nearest to two: zero, five, three, four, six, one, seven, nine
The nearest to which: to, is, a, the, that, it, and, with,
The nearest to seven: one, six, eight, five, nine, four, three, two,
The nearest to american: musician, actor, actress, nine, singer,
politician, d, one,
```

```
The nearest to many: often, as, most, modern, such, and, widely,
traditional,
The nearest to powers: constitutional, formally, power, rule, exercised,
parliamentary, constitution, control,
The nearest to question: questions, what, answer, existence, prove, merely,
true, statements,
The nearest to channel: network, channels, broadcasts, stations, cable,
broadcast, broadcasting, radio,
The nearest to recorded: songs, band, song, rock, album, bands, music,
studio,
The nearest to arts: art, school, martial, schools, students, styles,
education, student,
The nearest to orthodox: orthodoxy, churches, church, christianity,
christians, catholics, christian, oriental,
The nearest to scale: scales, can, amounts, depends, tend, are, structural,
for,
The nearest to mean: we, defined, is, exactly, equivalent, denote, number,
above,
Epoch Number 10/10 Iteration Number: 46100 Avg. Training loss: 4.8583
Epoch Number 10/10 Iteration Number: 46200 Avg. Training loss: 4.8887
```

As you can see from the output, the network somehow learned some semantically useful representation of the input words. To help us get a clearer picture of the embedding matrix, we are going to use a dimensionality reduction technique such as t-SNE to reduce the real-valued vectors to two dimensions, and then we'll visualize them and label each point with its corresponding word:

```
num_visualize_words = 500
tsne_obj = TSNE()
embedding_tsne =
tsne_obj.fit_transform(embedding_matrix[:num_visualize_words, :])

fig, ax = plt.subplots(figsize=(14, 14))
for ind in range(num_visualize_words):
    plt.scatter(*embedding_tsne[ind, :], color='steelblue')
    plt.annotate(integer_to_vocab[ind], (embedding_tsne[ind, 0],
embedding_tsne[ind, 1]), alpha=0.7)
```

Output:

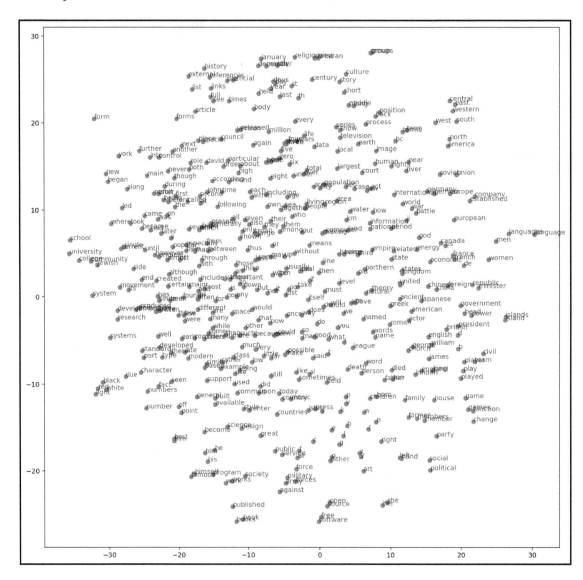

Figure 15.12: A visualization of word vectors

Summary

In this chapter, we went through the idea of representation learning and why it's useful for doing deep learning or machine learning in general on input that's not in a real-valued form. Also, we covered one of the adopted techniques for converting words into real-valued vectors—Word2Vec—which has very interesting properties. Finally, we implemented the Word2Vec model using the skip-gram architecture.

Next up, you will see the practical use of these learned representations in a sentiment analysis example, where we need to convert the input text to real-valued vectors.

12
Neural Sentiment Analysis

In this chapter, we are going to address one of the hot and trendy applications in natural language processing, which is called **sentiment analysis**. Most people nowadays express their opinions about something through social media platforms, and making use of this vast amount of text to keep track of customer satisfaction about something is very crucial for companies or even governments.

In this chapter, we are going to use recurrent-type neural networks to build a sentiment analysis solution. The following topics will be addressed in this chapter:

- General sentiment analysis architecture
- Sentiment analysis—model implementation

General sentiment analysis architecture

In this section, we are going to focus on the general deep learning architectures that can be used for sentiment analysis. The following figure shows the processing steps that are required for building the sentiment analysis model.

So, first off, we are going to deal with natural human language:

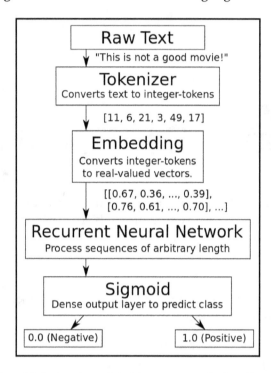

Figure 1: A general pipeline for sentiment analysis solutions or even sequence-based natural language solutions

We are going to use movie reviews to build this sentiment analysis application. The goal of this application is to produce positive and negative reviews based on the input raw text. For example, if the raw text is something like, **This movie is good**, then we need the model to produce a positive sentiment for it.

A sentiment analysis application will take us through a lot of processing steps that are needed to work with natural human languages inside a neural network such as embeddings.

So in this case, we have a raw text, for example, **This is not a good movie!** What we want to end up with is whether this is a negative or positive sentiment.

There are several difficulties in this type of application:

- One of them is that the sequences may have **different lengths**. This is a very short one, but we will see examples of text that have more than 500 words.
- Another problem is that if we just look at individual words (for example, good), that indicates a positive sentiment. However, it is preceded by the word **not**, so now it's a negative sentiment. This can get a lot more complicated, and we will see an example of it later.

As we learned in the previous chapter, a neural network cannot work on raw text, so we need to first convert it into what are called **tokens**. These are basically just integer values, so we go through our entire dataset and we count the number of times each word is being used. Then, we make a vocabulary and each word gets an index in this vocabulary. So the word **this** has an integer ID or token **11**, the word **is** has a token **6**, **not** has a token **21**, and so forth. So now, we have converted the raw text into a list of integers called tokens.

A neural network still cannot operate on this data, because if we have a vocabulary of 10,000 words, the tokens can take values between 0 and 9,999, and they may not be related at all. So, word number 998 may have a completely different semantic meaning than word number 999.

Therefore, we will use the idea of representation learning or embeddings that we learned about in the previous chapter. This embedding layer converts integer tokens into real-valued vectors, so token **11** becomes the vector [0.67,0.36,...,0.39], as shown in *Figure 1*. The same applies to the next token number 6.

A quick recap of what we studied in the previous chapter: this embedding layer in the preceding figure learns the mapping between tokens and their corresponding real-valued vector. Also, the embedding layer learns the semantic meanings of the words so that words that have similar meanings are somehow close to each other in this embedding space.

Out of the input raw text, we get a two-dimensional matrix, or tensor, which can now be inputted to the **recurrent neural network** (**RNN**). This can process sequences of arbitrary length and the output of this network is then fed into a fully connected or dense layer with a sigmoid activation function. So, the output is between 0 and 1, where a value of 0 is taken to mean a negative sentiment. But what if the value of the sigmoid function is neither 0 nor 1? Then we need to introduce a cut-off or a threshold value in the middle so that if the value is below 0.5, then the corresponding input is taken to be a negative sentiment, and a value above this threshold is taken to be a positive sentiment.

RNNs – sentiment analysis context

Now, let's recap the basic concepts of RNNs and also talk about them in the context of the sentiment analysis application. As we mentioned in the RNN chapter, the basic building block of a RNN is a recurrent unit, as shown in this figure:

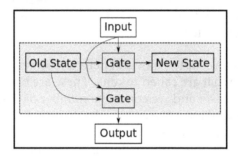

Figure 2: An abstract idea of an RNN unit

This figure is an abstraction of what goes on inside the recurrent unit. What we have here is the input, so this would be a word, for example, **good**. Of course, it has to be converted to embedding vectors. However, we will ignore that for now. Also, this unit has a kind of memory state, and depending on the contents of this **State** and the **Input**, we will update this state and write new data into the state. For example, imagine that we have previously seen the word **not** in the input; we write that to the state so that when we see the word **good** on one of the following inputs, we know from the state that we have just seen the word **not**. Now, we see the word **good**. Thus, we have to write into the state that we have seen the words **not good** together so that this might indicate that the whole input text probably has a negative sentiment.

The mapping from the old state and the input to the new contents of the state is done through a so-called **Gate**, and the way these are implemented differs across different versions of recurrent units. It is basically a matrix operation with an activation function, but as we will see in a moment, there is a problem with backpropagating gradients. So, the RNN has to be designed in a special way so that the gradients are not distorted too much.

In a recurrent unit, we have a similar gate for producing the output, and once again the output of the recurrent unit depends on the current contents of the state and the input that we are seeing. So what we can try and do is unroll the processing that takes place with a recurrent unit:

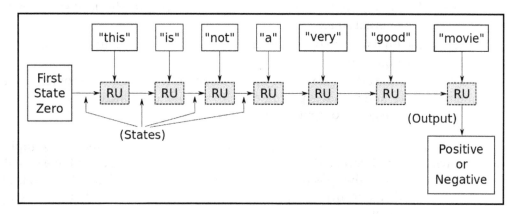

Figure 3: Unrolled version of the recurrent neural net

Now, what we have here is just one recurrent unit, but the flow chart shows what happens at different time steps. So:

- In time step 1, we input the word **this** to the recurrent unit and it has its internal memory state first initialized to zero. This is done by TensorFlow whenever we start processing a new sequence of data. So, we see the word **this** and the recurrent unit state is 0. Hence, we use the internal gate to update the memory state and **this** is then used in time step number two where we input the word **is**; now, the memory state has some contents. There's not a whole lot of meaning in the word **this**, so the state might still be around 0.
- And there's also not a lot of meaning in **is**, so perhaps the state is still somewhat 0.
- In the next time step, we see the word **not**, and this has meaning we ultimately want to predict, which is the sentiment of the whole input text. This one is what we need to store in the memory so that the gate inside the recurrent unit sees that the state already probably contains near-zero values. But now it wants to store what we have just seen the word **not**, so it saves some nonzero value in this state.
- Then, we move on to the next time step, where we have the word **a**; this also doesn't have much information, so it's probably just ignored. It just copies over the state.

- Now, we have the word **very**, and this indicates that whatever sentiment exists might be a strong sentiment, so the recurrent unit now knows that we have seen **not** and **very**. It stores this somehow in its memory state.
- In the next time step, we see the word **good**, so now the network knows **not very good** and it thinks, *Oh, this is probably a negative sentiment!* Hence, it stores that value in the internal state.
- Then, in the final time step, we see **movie**, and this is not really relevant, so it's probably just ignored.
- Next, we use the other gate inside the recurrent unit to output the contents of the memory state, and then it is processed with the sigmoid function (which we don't show here). We get an output value between 0 and 1.

The idea then is that we want to train this network on many many thousands of examples of movie reviews from the Internet Movie database, where, for each input text, we give it the true sentiment value of either positive or negative. Then, we want TensorFlow to find out what the gates inside the recurrent unit should be so that they accurately map this input text to the correct sentiment:

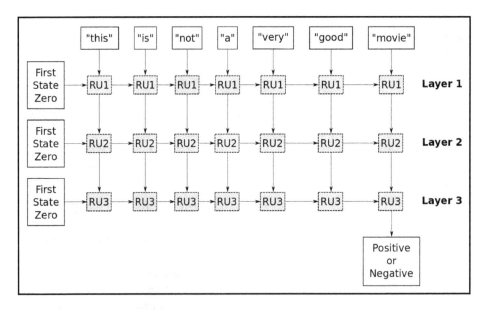

Figure 4: Used architecture for this chapter's implementation

The architecture for the RNN we will be using in this implementation is an RNN-type architecture with three layers. In the first layer, what we've just explained happens, except that now we need to output the value from the recurrent unit at each time step. Then, we gather a new sequence of data, which is the output of the first recurrent layer. Next, we can input it to the second recurrent layer because recurrent units need sequences of input data (and the output that we got from the first layer and the one that we want to feed into the second recurrent layer are some floating-point values whose meanings we don't really understand). This has a meaning inside the RNN, but it's not something we as humans will understand. Then, we do similar processing in the second recurrent layer.

So, first, we initialize the internal memory state of this recurrent unit to 0; then, we take the first output from the first recurrent layer and input it. We process it with the gates inside this recurrent unit, update the state, take the output of the first layer's recurrent unit for the second word **is**, and use that as input as well as the internal memory state. We continue doing this until we have processed the whole sequence, and then we gather up all the outputs of the second recurrent layer. We use them as inputs in the third recurrent layer, where we do a similar processing. But here, we only want the output for the last time step, which is a kind of summary for everything that has been fed so far. We then output that to a fully connected layer that we don't show here. Finally, we have the sigmoid activation function, so we get a value between zero and one, which represents negative and positive sentiments, respectively.

Exploding and vanishing gradients - recap

As we mentioned in the previous chapter, there's a phenomenon called **exploding** and **vanishing** of gradients values, which is very important in RNNs. Let's go back and look at *Figure 1*; that flowchart explains what this phenomenon is.

Imagine we have a text with 500 words in this dataset that we will be using to implement our sentiment analysis classifier. At every time step, we apply the internal gates in the recurrent unit in a recursive manner; so if there are 500 words, we will apply these gates 500 times to update the internal memory state of the recurrent unit.

As we know, the way neural networks are trained is by using so-called backpropagation of gradients, so we have some loss function that gets the output of the neural network and then the true output that we desire for the given input text. Then, we want to minimize this loss value so that the actual output of the neural network corresponds to the desired output for this particular input text. So, we need to take the gradient of this loss function with respect to the weights inside these recurrent units, and these weights are for the gates that are updating the internal state and outputting the value in the end.

Now, the gate is applied maybe 500 times, and if this has a multiplication in it, what we essentially get is an exponential function. So, if you multiply a value with itself 500 times and if this value is slightly less than 1, then it will very quickly vanish or get lost. Similarly, if a value slightly more than 1 is multiplied with itself 500 times, it'll explode.

The only values that can survive 500 multiplications are 0 and 1. They will remain the same, so the recurrent unit is actually much more complicated than what you see here. This is the abstract idea—that we want to somehow map the internal memory state and the input to update the internal memory state and to output some value—but in reality, we need to be very careful about propagating the gradients backwards through these gates so that we don't have this exponential multiplication over many many time steps. We also encourage you to see some tutorials on the mathematical definition of recurrent units.

Sentiment analysis – model implementation

We have seen all the bits and pieces of how to implement a stacked version of the LSTM variation of RNNs. To make things a bit exciting, we are going to use a higher level API called `Keras`.

Keras

> *"Keras is a high-level neural networks API, written in Python and capable of running on top of TensorFlow, CNTK, or Theano. It was developed with a focus on enabling fast experimentation. Being able to go from idea to result with the least possible delay is key to doing good research." – Keras website*

So, Keras is just a wrapper around TensorFlow and other deep learning frameworks. It's really good for prototyping and getting things built very quickly, but on the other hand, it gives you less control over your code. We'll take a chance to implement this sentiment analysis model in Keras so that you get a hands-on implementation in both TensorFlow and Keras. You can use Keras for fast prototyping and TensorFlow for a production-ready system.

More interesting news for you is that you don't have to switch to a totally different environment. You can now access Keras as a module in TensorFlow and import packages just like the following:

```
from tensorflow.python.keras.models
import Sequential
from tensorflow.python.keras.layers
import Dense, GRU, Embedding
from tensorflow.python.keras.optimizers
import Adam
from tensorflow.python.keras.preprocessing.text
import Tokenizer
from tensorflow.python.keras.preprocessing.sequence
import pad_sequences
```

So, let's go ahead and use what we can now call a more abstracted module inside TensorFlow that will help us to prototype deep learning solutions very fast. This is because we will get to write full deep learning solutions in just a few lines of code.

Data analysis and preprocessing

Now, let's move on to the actual implementation where we need to load the data. Keras actually has a functionality that can be used to load this sentiment dataset from IMDb, but the problem is that it has already mapped all the words to integer tokens. This is such an essential part of working with natural human language insight neural networks that I really want to show you how to do it.

Also, if you want to use this code for sentiment analysis of whatever data you might have in some other language, you will need to do this yourself, so we have just quickly implemented some functions for downloading this dataset.

Let's start off by importing a bunch of required packages:

```
%matplotlib inline
import matplotlib.pyplot as plt
import tensorflow as tf
import numpy as np
from scipy.spatial.distance import cdist
from tensorflow.python.keras.models import Sequential
from tensorflow.python.keras.layers import Dense, GRU, Embedding
from tensorflow.python.keras.optimizers import Adam
from tensorflow.python.keras.preprocessing.text import Tokenizer
from tensorflow.python.keras.preprocessing.sequence import pad_sequences
```

And then we load the dataset:

```
import imdb
imdb.maybe_download_and_extract()

Output:
- Download progress: 100.0%
Download finished. Extracting files.
Done.

input_text_train, target_train = imdb.load_data(train=True)
input_text_test, target_test = imdb.load_data(train=False)

print("Size of the trainig set: ", len(input_text_train))
print("Size of the testing set:  ", len(input_text_test))

Output:
Size of the trainig set: 25000
Size of the testing set: 25000
```

As you can see, it has 25,000 texts in the training set and in the testing set.

Let's just see one example from the training set and how it looks:

```
#combine dataset
text_data = input_text_train + input_text_test
input_text_train[1]

Output:
'This is a really heart-warming family movie. It has absolutely brilliant
animal training and "acting" (if you can call it like that) as well (just
think about the dog in "How the Grinch stole Christmas"... it was plain bad
training). The Paulie story is extremely well done, well reproduced and in
general the characters are really elaborated too. Not more to say except
that this is a GREAT MOVIE!<br /><br />My ratings: story 8.5/10, acting
7.5/10, animals+fx 8.5/10, cinematography 8/10.<br /><br />My overall
rating: 8/10 - BIG FAMILY MOVIE AND VERY WORTH WATCHING!'

target_train[1]

Output:
1.0
```

This is a fairly short one and the sentiment value is 1.0, which means it is a positive sentiment, so this is a positive review of whatever movie this was about.

Now, we get to the tokenizer, and this is the first step of processing this raw data because the neural network cannot work on text data. Keras has implemented what is called a **tokenizer** for building a vocabulary and mapping from words to an integer.

Also, we can say that we want a maximum of 10,000 words, so it will use only the 10,000 most popular words from the dataset:

```
num_top_words = 10000
tokenizer_obj = Tokenizer(num_words=num_top_words)
```

Now, we take all the text from the dataset and we call this function `fit` on texts:

```
tokenizer_obj.fit_on_texts(text_data)
```

The tokenizer takes about 10 seconds, and then it will have built the vocabulary. It looks like this:

```
tokenizer_obj.word_index

Output:
{'britains': 33206,
 'labcoats': 121364,
 'steeled': 102939,
 'geddon': 67551,
 "rossilini's": 91757,
 'recreational': 27654,
 'suffices': 43205,
 'hallelujah': 30337,
 'mallika': 30343,
 'kilogram': 122493,
 'elphic': 104809,
 'feebly': 32818,
 'unskillful': 91728,
 "'mistress'": 122218,
 "yesterday's": 25908,
 'busco': 85664,
 'goobacks': 85670,
 'mcfeast': 71175,
 'tamsin': 77763,
 "petron's": 72628,
 "'lion": 87485,
 'sams': 58341,
 'unbidden': 60042,
 "principal's": 44902,
 'minutiae': 31453,
 'smelled': 35009,
 'history\x97but': 75538,
```

```
    'vehemently': 28626,
    'leering': 14905,
    'kýnay': 107654,
    'intendend': 101260,
    'chomping': 21885,
    'nietsze': 76308,
    'browned': 83646,
    'grosse': 17645,
    "''gaslight''": 74713,
    'forseeing': 103637,
    'asteroids': 30997,
    'peevish': 49633,
    "attic'": 120936,
    'genres': 4026,
    'breckinridge': 17499,
    'wrist': 13996,
    "sopranos'": 50345,
    'embarasing': 92679,
    "wednesday's": 118413,
    'cervi': 39092,
    'felicity': 21570,
    "''horror''": 56254,
    'alarms': 17764,
    "'ol": 29410,
    'leper': 27793,
    'once\x85': 100641,
    'iverson': 66834,
    'triply': 117589,
    'industries': 19176,
    'brite': 16733,
    'amateur': 2459,
    "libby's": 46942,
    'eeeeevil': 120413,
    'jbc33': 51111,
    'wyoming': 12030,
    'waned': 30059,
    'uchida': 63203,
    'uttter': 93299,
    'irector': 123847,
    'outriders': 95156,
    'perd': 118465,
    .
    .
    .}
```

So, each word is now associated with an integer; therefore, the word `the` has number 1:

```
tokenizer_obj.word_index['the']

Output:
1
```

Here, `and` has number 2:

```
tokenizer_obj.word_index['and']

Output:
2
```

The word `a` has 3:

```
tokenizer_obj.word_index['a']

Output:
3
```

And so on. We see that `movie` has number 17:

```
tokenizer_obj.word_index['movie']

Output:
17
```

And `film` has number 19:

```
tokenizer_obj.word_index['film']

Output:
19
```

What all this means is that `the` was the most used word in the dataset and `and` was the second most used in the dataset. So, whenever we want to map words to integer tokens, we will get these numbers.

Let's try and take the word number 743 for example, and this was the word `romantic`:

```
tokenizer_obj.word_index['romantic']

Output:
743
```

So, whenever we see the word `romantic` in the input text, we map it to the token integer `743`. We use the tokenizer again to convert all the words in the text in the training set into integer tokens:

```
input_text_train[1]
Output:
'This is a really heart-warming family movie. It has absolutely brilliant
animal training and "acting" (if you can call it like that) as well (just
think about the dog in "How the Grinch stole Christmas"... it was plain bad
training). The Paulie story is extremely well done, well reproduced and in
general the characters are really elaborated too. Not more to say except
that this is a GREAT MOVIE!<br /><br />My ratings: story 8.5/10, acting
7.5/10, animals+fx 8.5/10, cinematography 8/10.<br /><br />My overall
rating: 8/10 - BIG FAMILY MOVIE AND VERY WORTH WATCHING!
```

When we convert that text to integer tokens, it becomes an array of integers:

```
np.array(input_train_tokens[1])

Output:
array([ 11,  6,  3,  62,  488,  4679,  236,  17,  9,  45,  419,
        513,  1717,  2425,  2,  113,  43,  22,  67,  654,  9,  37,
        12,  14,  69,  39,  101,  42,  1,  826,  8,  85,  1,
       6418,  3492,  1156,  9,  13,  1042,  74,  2425,  1,  6419,  64,
        6,  568,  69,  221,  69,  2,  8,  825,  1,  102,  23,
        62,  96,  21,  51,  5,  131,  556,  12,  11,  6,  3,
        78,  17,  7,  7,  56,  2818,  64,  723,  447,  156,  113,
       702,  447,  156,  1598,  3611,  723,  447,  156,  633,  723,  156,
        7,  7,  56,  437,  670,  723,  156,  191,  236,  17,  2,
        52,  278,  147])
```

So, the word `this` becomes the number 11, the word `is` becomes the number 59, and so forth.

We also need to convert the rest of the text:

```
input_test_tokens = tokenizer_obj.texts_to_sequences(input_text_test)
```

Now, there's another problem because the sequences of tokens have different lengths depending on the length of the original text, even though the recurrent units can work with sequences of arbitrary length. But the way that TensorFlow works is that all of the data in a batch needs to have the same length.

So, we can either ensure that all sequences in the entire dataset have the same length, or write a custom data generator that ensures that the sequences in a single batch have the same length. Now, it is a lot simpler to ensure that all the sequences in the dataset have the same length, but the problem is that there are some outliers. We have some sentences that, I think, are more than 2,200 words long. It will hurt our memory very much if we have all the *short* sentences with more than 2,200 words. So what we will do instead is make a compromise; first, we need to count all the words, or the number of tokens in each of these input sequences. What we see is that the average number of words in a sequence is about 221:

```
total_num_tokens = [len(tokens) for tokens in input_train_tokens +
input_test_tokens]
total_num_tokens = np.array(total_num_tokens)

#Get the average number of tokens
np.mean(total_num_tokens)

Output:
221.27716
```

And we see that the maximum number of words is more than 2,200:

```
np.max(total_num_tokens)

Output:
2208
```

Now, there's a huge difference between the average and the max, and again we would be wasting a lot of memory if we just padded all the sentences in the dataset so that they would all have `2208` tokens. This would especially be a problem if you have a dataset with millions of sequences of text.

So what we will do is make a compromise where we will pad all sequences and truncate the ones that are too long so that they have `544` words. The way we calculated this was like this—we took the average number of words in all the sequences in the dataset and we added two standard deviations:

```
max_num_tokens = np.mean(total_num_tokens) + 2 * np.std(total_num_tokens)
max_num_tokens = int(max_num_tokens)
max_num_tokens

Output:
544
```

What do we get out of this is? We cover about 95% of the text in the dataset, so only about 5% are longer than `544` words:

```
np.sum(total_num_tokens < max_num_tokens) / len(total_num_tokens)

Output:
0.94532
```

Now, we call these functions in Keras. They will either pad the sequences that are too short (so they will just add zeros) or truncate the sequences that are too long (basically just cut off some of the words if the text is too long).

Now, there's an important thing here: we can do this padding and truncating in pre or post mode. So imagine we have a sequence of integer tokens and we want to pad this because it's too short. We can:

- Either pad all of these zeros at the beginning so that we have the actual integer tokens down at the end.
- Or do it in the opposite way so that we have all this data at the beginning and then all the zeros at the end. But if we just go back and look at the preceding RNN flowchart, remember that it is processing the sequence one step at a time so if we start processing zeros, it will probably not mean anything and the internal state would have probably just remain zero. So, whenever it finally sees an integer token for a specific word, it will know okay now we start processing the data.

However, if all the zeros were at the end, we would have started processing all the data; then we'd have some internal state inside the recurrent unit. Right now, we see a whole lot of zeros, so that might actually destroy the internal state that we have just calculated. This is why it might be a good idea to have the zeros padded at the beginning.

But the other problem is when we truncate a text, so if the text is very long, we will truncate it to get it to fit to `544` words, or whatever the number was. Now, imagine we've caught this sentence here in the middle somewhere and it says **this very good movie** or **this is not**. You know, of course, that we do this only for very long sequences, but it is possible that we lose essential information for properly classifying this text. So it is a compromise that we're making when we are truncating input text. A better way would be to create a batch and just pad text inside that batch. So, when we see a very very long sequence, we pad the other sequences to have the same length. But we don't need to store all of this data in memory because most of it is wasted.

Let's go back and convert the entire dataset so that it is truncated and padded; thus, it's one big matrix of data:

```
seq_pad = 'pre'

input_train_pad = pad_sequences(input_train_tokens, maxlen=max_num_tokens,
  padding=seq_pad, truncating=seq_pad)

input_test_pad = pad_sequences(input_test_tokens, maxlen=max_num_tokens,
  padding=seq_pad, truncating=seq_pad)
```

We check the shape of this matrix:

```
input_train_pad.shape

Output:
(25000, 544)

input_test_pad.shape

Output:
(25000, 544)
```

So, let's have a look at specific sample tokens before and after padding:

```
np.array(input_train_tokens[1])

Output:
array([ 11,  6,  3,  62,  488,  4679,  236,  17,  9,  45,  419,
        513,  1717,  2425,  2,  113,  43,  22,  67,  654,  9,  37,
        12,  14,  69,  39,  101,  42,  1,  826,  8,  85,  1,
      6418,  3492,  1156,  9,  13,  1042,  74,  2425,  1,  6419,  64,
        6,  568,  69,  221,  69,  2,  8,  825,  1,  102,  23,
        62,  96,  21,  51,  5,  131,  556,  12,  11,  6,  3,
        78,  17,  7,  7,  56,  2818,  64,  723,  447,  156,  113,
       702,  447,  156,  1598,  3611,  723,  447,  156,  633,  723,  156,
        7,  7,  56,  437,  670,  723,  156,  191,  236,  17,  2,
        52,  278,  147])
```

And after padding, this sample will look like the following:

```
input_train_pad[1]

Output:
array([ 0,  0,  0,  0,  0,  0,  0,  0,  0,  0,  0,
        0,  0,  0,  0,  0,  0,  0,  0,  0,  0,  0,
        0,  0,  0,  0,  0,  0,  0,  0,  0,  0,  0,
        0,  0,  0,  0,  0,  0,  0,  0,  0,  0,  0,
```

```
           0,  0,  0,  0,  0,  0,  0,  0,  0,  0,  0,
           0,  0,  0,  0,  0,  0,  0,  0,  0,  0,  0,
           0,  0,  0,  0,  0,  0,  0,  0,  0,  0,  0,
           0,  0,  0,  0,  0,  0,  0,  0,  0,  0,  0,
           0,  0,  0,  0,  0,  0,  0,  0,  0,  0,  0,
           0,  0,  0,  0,  0,  0,  0,  0,  0,  0,  0,
           0,  0,  0,  0,  0,  0,  0,  0,  0,  0,  0,
           0,  0,  0,  0,  0,  0,  0,  0,  0,  0,  0,
           0,  0,  0,  0,  0,  0,  0,  0,  0,  0,  0,
           0,  0,  0,  0,  0,  0,  0,  0,  0,  0,  0,
           0,  0,  0,  0,  0,  0,  0,  0,  0,  0,  0,
           0,  0,  0,  0,  0,  0,  0,  0,  0,  0,  0,
           0,  0,  0,  0,  0,  0,  0,  0,  0,  0,  0,
           0,  0,  0,  0,  0,  0,  0,  0,  0,  0,  0,
           0,  0,  0,  0,  0,  0,  0,  0,  0,  0,  0,
           0,  0,  0,  0,  0,  0,  0,  0,  0,  0,  0,
           0,  0,  0,  0,  0,  0,  0,  0,  0,  0,  0,
           0,  0,  0,  0,  0,  0,  0,  0,  0,  0,  0,
           0,  0,  0,  0,  0,  0,  0,  0,  0,  0,  0,
           0,  0,  0,  0,  0,  0,  0,  0,  0,  0,  0,
           0,  0,  0,  0,  0,  0,  0,  0,  0,  0,  0,
           0,  0,  0,  0,  0,  0,  0,  0,  0,  0,  0,
           0,  0,  0,  0,  0,  0,  0,  0,  0,  0,  0,
           0,  0,  0,  0,  0,  0,  0,  0,  0,  0,  0,
           0,  0,  0,  0,  0,  0,  0,  0,  0,  0,  0,
           0,  0,  0,  0,  0,  0,  0,  0,  0,  0,  0,
           0,  0,  0,  0,  0,  0,  0,  0,  0,  0,  0,
           0,  0,  0,  0,  0,  0,  0,  0,  0,  0,  0,
           0,  0,  0,  0,  0,  0,  0,  0,  0,  0,  0,
           0,  0,  0,  0,  0,  0,  0,  0,  0,  0,  0,
           0,  0,  0,  0,  0,  0,  0,  0,  0,  0,  0,
           0,  0,  0,  0, 0,  0,  0,  0,  0,  0,  0,
           0,  0,  0,  0,  0,  0,  0,  0,  0,  0,  0,
           0,  0,  0,  0,  0,  0,  0,  0,  0,  0,  0,
           0,  0,  0,  0,  0,  0,  0,  0,  0,  0,  0,
           0,  0,  0,  0,  0,  0,  0,  0,  0,  0,  0,
           0,  0, 11,  6,  3, 62, 488, 4679, 236, 17, 9,
          45, 419, 513, 1717, 2425, 2, 113, 43, 22, 67, 654,
           9, 37, 12, 14, 69, 39, 101, 42, 1, 826, 8,
          85,  1, 6418, 3492, 1156, 9, 13, 1042, 74, 2425, 1,
        6419, 64,  6, 568, 69, 221, 69, 2, 8, 825, 1,
         102, 23, 62, 96, 21, 51, 5, 131, 556, 12, 11,
           6,  3, 78, 17,  7,  7, 56, 2818, 64, 723, 447,
         156, 113, 702, 447, 156, 1598, 3611, 723, 447, 156, 633,
         723, 156,  7,  7, 56, 437, 670, 723, 156, 191, 236,
          17,  2, 52, 278, 147], dtype=int32)
```

Also, we need a functionality to map backwards so that it maps from integer tokens back to text words; we just need that here. It's a very simple helper function, so let's go ahead and implement it:

```
index = tokenizer_obj.word_index
index_inverse_map = dict(zip(index.values(), index.keys()))

def convert_tokens_to_string(input_tokens):

  # Convert the tokens back to words
  input_words = [index_inverse_map[token] for token in input_tokens if token
!= 0]

  # join them all words.
  combined_text = " ".join(input_words)

return combined_text
```

Now, for example, the original text in the dataset is like this:

```
input_text_train[1]
Output:

input_text_train[1]

'This is a really heart-warming family movie. It has absolutely brilliant
animal training and "acting" (if you can call it like that) as well (just
think about the dog in "How the Grinch stole Christmas"... it was plain bad
training). The Paulie story is extremely well done, well reproduced and in
general the characters are really elaborated too. Not more to say except
that this is a GREAT MOVIE!<br /><br />My ratings: story 8.5/10, acting
7.5/10, animals+fx 8.5/10, cinematography 8/10.<br /><br />My overall
rating: 8/10 - BIG FAMILY MOVIE AND VERY WORTH WATCHING!'
```

If we use a helper function to convert the tokens back to text words, we get this text:

```
convert_tokens_to_string(input_train_tokens[1])

'this is a really heart warming family movie it has absolutely brilliant
animal training and acting if you can call it like that as well just think
about the dog in how the grinch stole christmas it was plain bad training
the paulie story is extremely well done well and in general the characters
are really too not more to say except that this is a great movie br br my
ratings story 8 5 10 acting 7 5 10 animals fx 8 5 10 cinematography 8 10 br
br my overall rating 8 10 big family movie and very worth watching'
```

It's basically the same except for punctuation and other symbols.

Building the model

Now, we need to create the RNN, and we will do this in Keras because it's very simple. We do that with the so-called `sequential` model.

The first layer of this architecture will be what is called an **embedding**. If we look back at the flowchart in *Figure 1*, what we just did was we converted the raw input text to integer tokens. But we still cannot input this to a RNN, so we have to convert that into embedding vectors, which are values that are somewhere between -1 and 1. They can exceed to some extent but are generally somewhere between -1 and 1, and this is data that we can then work on in the neural network.

It's somewhat magical because this embedding layer trains simultaneously with the RNN and it doesn't see the raw words. It sees integer tokens but learns to recognize that there is some pattern in how words are being used together. So it can, sort of, deduce that some words or some integer tokens have similar meaning, and then it encodes this in embedding vectors that look somewhat the same.

Therefore, what we need to decide is the length of each vector so that, for example, the token "11" gets converted into a real-valued vector. In this example, we will use a length of 8, which is actually extremely short (normally, it is somewhere between 100 and 300). Try and change this number of elements in the embedding vectors and rerun this code to see what you get as a result.

So, we set the embedding size to 8 and then use Keras to add this embedding layer to the RNN. This has to be the first layer in the network:

```
embedding_layer_size = 8

rnn_type_model.add(Embedding(input_dim=num_top_words,
                   output_dim=embedding_layer_size,
                   input_length=max_num_tokens,
                   name='embedding_layer'))
```

Then, we can add the first recurrent layer, and we will use what is called a **Gated Recurrent Unit (GRU)**. Often, you will see that people use what is called **LSTM**, but others seem to suggest that the GRU is better because there are gates inside LSTM that are redundant. And indeed the simpler code works just as well with fewer gates. You could add a thousand more gates to LSTM and that still doesn't mean it gets better.

So, let's define our GRU architectures; we say that we want an output dimensionality of 16 and we need to return sequences:

```
rnn_type_model.add(GRU(units=16, return_sequences=True))
```

If we look at the flowchart in *Figure 4*, we want to add a second recurrent layer:

```
rnn_type_model.add(GRU(units=8, return_sequences=True))
```

Then, we have the third and final recurrent layer, which will not output a sequence because it will be followed by a dense layer; it should only give the final output of the GRU and not a whole sequence of outputs:

```
rnn_type_model.add(GRU(units=4))
```

Then, the output here will be fed into a fully connected or dense layer, which is just supposed to output one value for each input sequence. This is processed with the sigmoid activation function so it outputs a value between 0 and 1:

```
rnn_type_model.add(Dense(1, activation='sigmoid'))
```

Then, we say we want to use the Adam optimizer with this learning rate here, and the loss function should be the binary cross-entropy between the output from the RNN and the actual class value from the training set, which will be a value of either 0 or 1:

```
model_optimizer = Adam(lr=1e-3)

rnn_type_model.compile(loss='binary_crossentropy',
            optimizer=model_optimizer,
            metrics=['accuracy'])
```

And now, we can just print a summary of what the model looks like:

```
rnn_type_model.summary()
```

```
Layer (type) Output Shape Param #
=================================================================
embedding_layer (Embedding) (None, 544, 8) 80000

gru_1 (GRU) (None, None, 16) 1200

gru_2 (GRU) (None, None, 8) 600

gru_3 (GRU) (None, 4) 156

dense_1 (Dense) (None, 1) 5
=================================================================
Total params: 81,961
Trainable params: 81,961
Non-trainable params: 0
```

So, as you can see, we have the embedding layer, the first recurrent unit, the second, third, and dense layer. Note that this doesn't have a lot of parameters.

Model training and results analysis

Now, it's time to kick off the training process, which is very easy here:

```
Output:
rnn_type_model.fit(input_train_pad, target_train,
        validation_split=0.05, epochs=3, batch_size=64)
```

```
Output:
Train on 23750 samples, validate on 1250 samples
Epoch 1/3
23750/23750 [==============================]23750/23750
[==============================] - 176s 7ms/step - loss: 0.6698 - acc:
0.5758 - val_loss: 0.5039 - val_acc: 0.7784

Epoch 2/3
23750/23750 [==============================]23750/23750
[==============================] - 175s 7ms/step - loss: 0.4631 - acc:
0.7834 - val_loss: 0.2571 - val_acc: 0.8960

Epoch 3/3
23750/23750 [==============================]23750/23750
[==============================] - 174s 7ms/step - loss: 0.3256 - acc:
0.8673 - val_loss: 0.3266 - val_acc: 0.8600
```

Let's test the trained model against the test set:

```
model_result = rnn_type_model.evaluate(input_test_pad, target_test)
```

```
Output:
25000/25000 [==============================]25000/25000
[==============================] - 60s 2ms/step
```

```
print("Accuracy: {0:.2%}".format(model_result[1]))
Output:
Accuracy: 85.26%
```

Now, let's see an example of some misclassified texts.

So first, we calculate the predicted classes for the first 1,000 sequences in the test set and then we take the actual class values. We compare them and get a list of indices where this mismatch exists:

```
target_predicted = rnn_type_model.predict(x=input_test_pad[0:1000])
target_predicted = target_predicted.T[0]
```

Use the cut-off threshold to indicate that all values above 0.5 will be considered positive and the others will be considered negative:

```
class_predicted = np.array([1.0 if prob>0.5 else 0.0 for prob in
target_predicted])
```

Now, let's get the actual class for these 1,000 sequences:

```
class_actual = np.array(target_test[0:1000])
```

Let's get the incorrect samples from the output:

```
incorrect_samples = np.where(class_predicted != class_actual)
incorrect_samples = incorrect_samples[0]
len(incorrect_samples)

Output:
122
```

So, we see that there are 122 of these texts that were incorrectly classified; that's 12.1% of the 1,000 texts we calculated here. Let's look at the first misclassified text:

```
index = incorrect_samples[0]
index

Output:
9

incorrectly_predicted_text = input_text_test[index]
incorrectly_predicted_text

Output:

'I am not a big music video fan. I think music videos take away personal
feelings about a particular song.. Any song. In other words, creative
thinking goes out the window. Likewise, Personal feelings aside about MJ,
toss aside. This was the best music video of alltime. Simply wonderful. It
was a movie. Yes folks it was. Brilliant! You had awesome acting, awesome
choreography, and awesome singing. This was spectacular. Simply a plot line
of a beautiful young lady dating a man, but was he a man or something
sinister. Vincent Price did his thing adding to the song and video. MJ was
```

MJ, enough said about that. This song was to video, what Jaguars are for cars. Top of the line, PERFECTO. What was even better about this was, that we got the real MJ without the thousand facelifts. Though ironically enough, there was more than enough makeup and costumes to go around. Folks go to Youtube. Take 14 mins. out of your life and see for yourself what a wonderful work of art this particular video really is.'

Let's have a look at the model output for this sample as well as the actual class:

```
target_predicted[index]
```

```
Output:
0.1529513
```

```
class_actual[index]
Output:
1.0
```

Now, let's test our trained model against a set of new data samples and see its results:

```
test_sample_1 = "This movie is fantastic! I really like it because it is so good!"
test_sample_2 = "Good movie!"
test_sample_3 = "Maybe I like this movie."
test_sample_4 = "Meh ..."
test_sample_5 = "If I were a drunk teenager then this movie might be good."
test_sample_6 = "Bad movie!"
test_sample_7 = "Not a good movie!"
test_sample_8 = "This movie really sucks! Can I get my money back please?"
test_samples = [test_sample_1, test_sample_2, test_sample_3, test_sample_4, test_sample_5, test_sample_6, test_sample_7, test_sample_8]
```

Now, let's convert them to integer tokens:

```
test_samples_tokens = tokenizer_obj.texts_to_sequences(test_samples)
```

And then pad them:

```
test_samples_tokens_pad = pad_sequences(test_samples_tokens, maxlen=max_num_tokens,
                              padding=seq_pad, truncating=seq_pad)
test_samples_tokens_pad.shape
```

```
Output:
(8, 544)
```

Finally, let's run the model against them:

```
rnn_type_model.predict(test_samples_tokens_pad)

Output:
array([[0.9496784 ],
  [0.9552593 ],
  [0.9115685 ],
  [0.9464672 ],
  [0.87672734],
  [0.81883633],
  [0.33248223],
  [0.15345531 ]], dtype=float32)
```

So, a value close to zero means a negative sentiment and a value that's close to 1 means a positive sentiment; finally, these numbers will vary every time you train the model.

Summary

In this chapter, we covered an interesting application, which is sentiment analysis. Sentiment analysis is used by different companies to track customer's satisfaction with their products. Even governments use sentiment analysis solutions to track citizen satisfaction about something that they want to do in the future.

Next up, we are going to focus on some advanced deep learning architectures that can be used for semi-supervised and unsupervised applications.

13
Autoencoders – Feature Extraction and Denoising

An autoencoder network is nowadays one of the widely used deep learning architectures. It's mainly used for unsupervised learning of efficient decoding tasks. It can also be used for dimensionality reduction by learning an encoding or a representation for a specific dataset. Using autoencoders in this chapter, we'll show how to denoise your dataset by constructing another dataset with the same dimensions but less noise. To use this concept in practice, we will extract the important features from the MNIST dataset and try to see how the performance will be significantly enhanced by this.

The following topics will be covered in this chapter:

- Introduction to autoencoders
- Examples of autoencoders
- Autoencoder architectures
- Compressing the MNIST dataset
- Convolutional autoencoders
- Denoising autoencoders
- Applications of autoencoders

Introduction to autoencoders

An autoencoder is yet another deep learning architecture that can be used for many interesting tasks, but it can also be considered as a variation of the vanilla feed-forward neural network, where the output has the same dimensions as the input. As shown in *Figure 1*, the way autoencoders work is by feeding data samples $(x_1,...,x_6)$ to the network. It will try to learn a lower representation of this data in layer *L2*, which you might call a way of encoding your dataset in a lower representation. Then, the second part of the network, which you might call a decoder, is responsible for constructing an output from this representation $(\hat{x}_1,\cdots,\hat{x}_6)$. You can think of the intermediate lower representation that the network learns from the input data as a compressed version of it.

Not very different from all the other deep learning architectures that we have seen so far, autoencoders use backpropagation algorithms.

An autoencoder neural network is an unsupervised learning algorithm that applies backpropagation, setting the target values to be equal to the inputs:

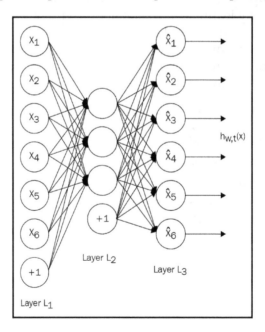

Figure 1: General autoencoder architecture

Examples of autoencoders

In this chapter, we will demonstrate some examples of different variations of autoencoders using the MNIST dataset. As a concrete example, suppose the inputs x are the pixel intensity values from a 28 x 28 image (784 pixels); so the number of input data samples is $n=784$. There are $s2=392$ hidden units in layer $L2$. And since the output will be of the same dimensions as the input data samples, $y \in R784$. The number of neurons in the input layer will be 784, followed by 392 neurons in the middle layer $L2$; so the network will be a lower representation, which is a compressed version of the output. The network will then feed this compressed lower representation of the input $a(L2) \in R392$ to the second part of the network, which will try hard to reconstruct the input pixels 784 from this compressed version.

Autoencoders rely on the fact that the input samples represented by the image pixels will be somehow correlated and then it will use this fact to reconstruct them. So autoencoders are a bit similar to dimensionality reduction techniques, because they learn a lower representation of the input data as well.

To sum up, a typical autoencoder will consist of three parts:

1. The encoder part, which is responsible for compressing the input into a lower representation
2. The code, which is the intermediate result of the encoder
3. The decoder, which is responsible for reconstructing the the original input using this code

The following figure shows the three main components of a typical autoencoder:

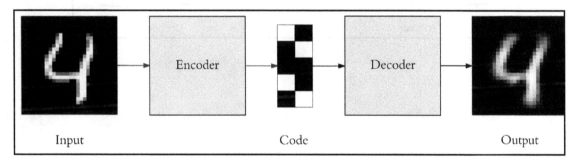

Figure 2: How encoders function over an image

As we mentioned, autoencoders part learn a compressed representation of the input that are then fed to the third part, which tries to reconstruct the input. The reconstructed input will be similar to the output but it won't be exactly the same as the original output, so autoencoders can't be used for compression tasks.

Autoencoder architectures

As we mentioned, a typical autoencoder consists of three parts. Let's explore these three parts in more detail. To motivate you, we are not going to reinvent the wheel here in this chapter. The encoder-decoder part is nothing but a fully connected neural network, and the code part is another neural network but it's not fully connected. The dimensionality of this code part is controllable and we can treat it as a hyperparameter:

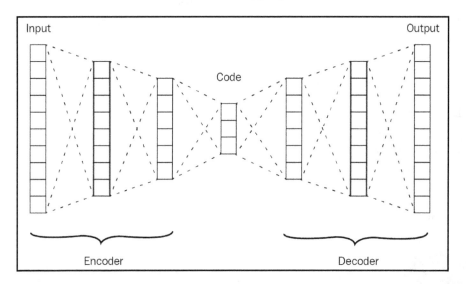

Figure 3: General encoder-decoder architecture of autoencoders

Before diving into using autoencoders for compressing the MNIST dataset, we are going to list the set of hyperparameters that we can use to fine-tune the autoencoder model. There are mainly four hyperparameters:

1. **Code part size**: This is the number of units in the middle layer. The lower the number of units we have in this layer, the more compressed the representation of the input we get.
2. **Number of layers in the encoder and decoder**: As we mentioned, the encoder and decoder are nothing but a fully connected neural network that we can make as deep as we can by adding more layers.
3. **Number of units per layer**: We can also use a different number of units in each layer. The shape of the encoder and decoder is very similar to DeconvNets, where the number of layers in the encoders decreases as we approach the code part and then starts to increase as we approach the final layer of the decoder.
4. **Model loss function**: We can use different loss functions as well, such as MSE or cross-entropy.

After defining these hyperparameters and giving them initial values, we can train the network using a backpropagation algorithm.

Compressing the MNIST dataset

In this part, we'll build a simple autoencoder that can be used to compress the MNIST dataset. So we will feed the images of this dataset to the encoder part, which will try to learn a lower compressed representation for them; then we will try to construct the input images again in the decoder part.

The MNIST dataset

We will start the implementation by getting the MNIST dataset, using the helper functions of TensorFlow.

Let's import the necessary packages for this implementation:

```
%matplotlib inline

import numpy as np
import tensorflow as tf
import matplotlib.pyplot as plt
```

```
from tensorflow.examples.tutorials.mnist import input_data
mnist_dataset = input_data.read_data_sets('MNIST_data', validation_size=0)
```

```
Output:
Extracting MNIST_data/train-images-idx3-ubyte.gz
Extracting MNIST_data/train-labels-idx1-ubyte.gz
Extracting MNIST_data/t10k-images-idx3-ubyte.gz
Extracting MNIST_data/t10k-labels-idx1-ubyte.gz
```

Let's start off by plotting some examples from the MNIST dataset:

```
# Plotting one image from the training set.
image = mnist_dataset.train.images[2]
plt.imshow(image.reshape((28, 28)), cmap='Greys_r')
```

```
Output:
```

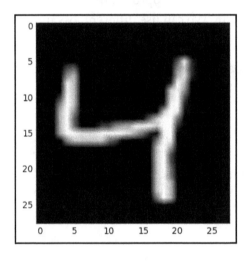

Figure 4: Example image from the MNIST dataset

```
# Plotting one image from the training set.
image = mnist_dataset.train.images[2]
plt.imshow(image.reshape((28, 28)), cmap='Greys_r')
```

```
Output:
```

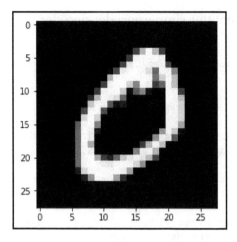

Figure 5: Example image from the MNIST dataset

Building the model

In order to build the encoder, we need to figure out how many pixels each MNIST image will have so that we can figure out the size of the input layer of the encoder. Each image from the MNIST dataset is 28 by 28 pixels, so we will reshape this matrix to a vector of 28 x 28 = 784 pixel values. We don't have to normalize the images of MNIST because they are already normalized.

Let's start off building our three components of the model. In this implementation, we will use a very simple architecture of a single hidden layer followed by ReLU activation, as shown in the following figure:

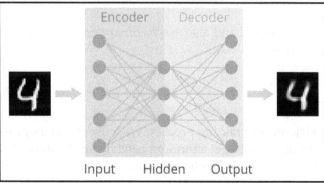

Figure 6: Encoder-decoder architecture for MNIST implementation

Let's go ahead and implement this simple encoder-decoder architecture according to the preceding explanation:

```
# The size of the encoding layer or the hidden layer.
encoding_layer_dim = 32

img_size = mnist_dataset.train.images.shape[1]

# defining placeholder variables of the input and target values
inputs_values = tf.placeholder(tf.float32, (None, img_size),
name="inputs_values")
targets_values = tf.placeholder(tf.float32, (None, img_size),
name="targets_values")

# Defining an encoding layer which takes the input values and incode them.
encoding_layer = tf.layers.dense(inputs_values, encoding_layer_dim,
activation=tf.nn.relu)

# Defining the logit layer, which is a fully-connected layer but without
any activation applied to its output
logits_layer = tf.layers.dense(encoding_layer, img_size, activation=None)

# Adding a sigmoid layer after the logit layer
decoding_layer = tf.sigmoid(logits_layer, name = "decoding_layer")

# use the sigmoid cross entropy as a loss function
model_loss = tf.nn.sigmoid_cross_entropy_with_logits(logits=logits_layer,
labels=targets_values)

# Averaging the loss values accross the input data
model_cost = tf.reduce_mean(model_loss)

# Now we have a cost functiont that we need to optimize using Adam
Optimizer
model_optimizier = tf.train.AdamOptimizer().minimize(model_cost)
```

Now we have defined our model and also used a binary cross-entropy since the images, pixels are already normalized.

Model training

In this section, we'll kick off the training process. We'll use the helper function of the `mnist_dataset` object in order to get a random batch from the dataset with a specific size; then we'll run the optimizer on this batch of images.

Let's start this section by creating the session variable, which will be responsible for executing the computational graph that we defined earlier:

```
# creating the session
 sess = tf.Session()
```

Next up, let's kick off the training process:

```
num_epochs = 20
train_batch_size = 200

sess.run(tf.global_variables_initializer())
for e in range(num_epochs):
    for ii in range(mnist_dataset.train.num_examples//train_batch_size):
        input_batch = mnist_dataset.train.next_batch(train_batch_size)
        feed_dict = {inputs_values: input_batch[0], targets_values:
input_batch[0]}
        input_batch_cost, _ = sess.run([model_cost, model_optimizier],
feed_dict=feed_dict)

        print("Epoch: {}/{}...".format(e+1, num_epochs),
            "Training loss: {:.3f}".format(input_batch_cost))
```

```
Output:
.
.
.
Epoch: 20/20... Training loss: 0.091
Epoch: 20/20... Training loss: 0.091
Epoch: 20/20... Training loss: 0.093
Epoch: 20/20... Training loss: 0.093
Epoch: 20/20... Training loss: 0.095
Epoch: 20/20... Training loss: 0.095
Epoch: 20/20... Training loss: 0.089
Epoch: 20/20... Training loss: 0.095
Epoch: 20/20... Training loss: 0.095
Epoch: 20/20... Training loss: 0.096
Epoch: 20/20... Training loss: 0.094
Epoch: 20/20... Training loss: 0.093
Epoch: 20/20... Training loss: 0.094
Epoch: 20/20... Training loss: 0.093
Epoch: 20/20... Training loss: 0.095
Epoch: 20/20... Training loss: 0.094
Epoch: 20/20... Training loss: 0.096
Epoch: 20/20... Training loss: 0.092
Epoch: 20/20... Training loss: 0.093
Epoch: 20/20... Training loss: 0.091
Epoch: 20/20... Training loss: 0.093
```

```
Epoch: 20/20... Training loss: 0.091
Epoch: 20/20... Training loss: 0.095
Epoch: 20/20... Training loss: 0.094
Epoch: 20/20... Training loss: 0.091
Epoch: 20/20... Training loss: 0.096
Epoch: 20/20... Training loss: 0.089
Epoch: 20/20... Training loss: 0.090
Epoch: 20/20... Training loss: 0.094
Epoch: 20/20... Training loss: 0.088
Epoch: 20/20... Training loss: 0.094
Epoch: 20/20... Training loss: 0.093
Epoch: 20/20... Training loss: 0.091
Epoch: 20/20... Training loss: 0.095
Epoch: 20/20... Training loss: 0.093
Epoch: 20/20... Training loss: 0.091
Epoch: 20/20... Training loss: 0.094
Epoch: 20/20... Training loss: 0.090
Epoch: 20/20... Training loss: 0.091
Epoch: 20/20... Training loss: 0.095
Epoch: 20/20... Training loss: 0.095
Epoch: 20/20... Training loss: 0.094
Epoch: 20/20... Training loss: 0.092
Epoch: 20/20... Training loss: 0.092
Epoch: 20/20... Training loss: 0.093
Epoch: 20/20... Training loss: 0.093
```

After running the preceding code snippet for 20 epochs, we will get a trained model that is able to generate or reconstruct images from the test set of the MNIST data. Bear in mind that if we feed images that are not similar to the ones that the model was trained on, then the reconstruction process just won't work because autoencoders are data-specific.

Let's test the trained model by feeding some images from the test set and see how the model is able to reconstruct them in the decoder part:

```
fig, axes = plt.subplots(nrows=2, ncols=10, sharex=True, sharey=True,
figsize=(20,4))

input_images = mnist_dataset.test.images[:10]
reconstructed_images, compressed_images = sess.run([decoding_layer,
encoding_layer], feed_dict={inputs_values: input_images})

for imgs, row in zip([input_images, reconstructed_images], axes):
    for img, ax in zip(imgs, row):
        ax.imshow(img.reshape((28, 28)), cmap='Greys_r')
        ax.get_xaxis().set_visible(False)
        ax.get_yaxis().set_visible(False)
```

```
fig.tight_layout(pad=0.1)
```

Output:

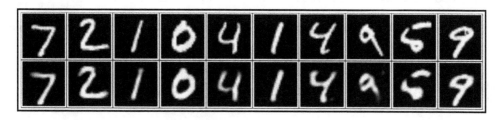

Figure 7: Examples of the original test images (first row) and their constructions (second row)

As you can see, the reconstructed images are very close to the input ones, but we can probably get better images using convolution layers in the encoder-decoder part.

Convolutional autoencoder

The previous simple implementation did a good job while trying to reconstruct input images from the MNIST dataset, but we can get a better performance through a convolution layer in the encoder and the decoder parts of the autoencoder. The resulting network of this replacement is called **convolutional autoencoder** (**CAE**). This flexibility of being able to replace layers is a great advantage of autoencoders and makes them applicable to different domains.

The architecture that we'll be using for the CAE will contain upsampling layers in the decoder part of the network to get the reconstructed version of the image.

Dataset

In this implementation, we can use any kind of imaging dataset and see how the convolutional version of the autoencoder will make a difference. We will still be using the MNIST dataset for this, so let's start off by getting the dataset using the TensorFlow helpers:

```
%matplotlib inline

import numpy as np
import tensorflow as tf
import matplotlib.pyplot as plt

from tensorflow.examples.tutorials.mnist import input_data
```

```
mnist_dataset = input_data.read_data_sets('MNIST_data', validation_size=0)

Output:
from tensorflow.examples.tutorials.mnist import input_data

mnist_dataset = input_data.read_data_sets('MNIST_data', validation_size=0)

Extracting MNIST_data/train-images-idx3-ubyte.gz
Extracting MNIST_data/train-labels-idx1-ubyte.gz
Extracting MNIST_data/t10k-images-idx3-ubyte.gz
Extracting MNIST_data/t10k-labels-idx1-ubyte.gz
```

Let's show one digit from the dataset:

```
# Plotting one image from the training set.
image = mnist_dataset.train.images[2]
plt.imshow(image.reshape((28, 28)), cmap='Greys_r')
```

Output:

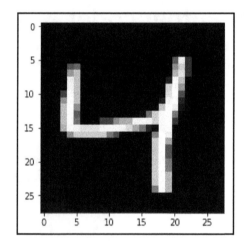

Figure 8: Example image from the MNIST dataset

Building the model

In this implementation, we will be using convolution layers with stride 1, and the padding parameter is set to be the same. By this, we won't change the height or width of the image. Also, we are using a set of max pooling layers to reduce the width and height of the image and hence building a compressed lower representation of the image.

So let's go ahead and build the core of our network:

```
learning_rate = 0.001

# Define the placeholder variable sfor the input and target values
inputs_values = tf.placeholder(tf.float32, (None, 28,28,1),
name="inputs_values")
targets_values = tf.placeholder(tf.float32, (None, 28,28,1),
name="targets_values")

# Defining the Encoder part of the netowrk
# Defining the first convolution layer in the encoder parrt
# The output tenosor will be in the shape of 28x28x16
conv_layer_1 = tf.layers.conv2d(inputs=inputs_values, filters=16,
kernel_size=(3,3), padding='same', activation=tf.nn.relu)

# The output tenosor will be in the shape of 14x14x16
maxpool_layer_1 = tf.layers.max_pooling2d(conv_layer_1, pool_size=(2,2),
strides=(2,2), padding='same')

# The output tenosor will be in the shape of 14x14x8
conv_layer_2 = tf.layers.conv2d(inputs=maxpool_layer_1, filters=8,
kernel_size=(3,3), padding='same', activation=tf.nn.relu)

# The output tenosor will be in the shape of 7x7x8
maxpool_layer_2 = tf.layers.max_pooling2d(conv_layer_2, pool_size=(2,2),
strides=(2,2), padding='same')

# The output tenosor will be in the shape of 7x7x8
conv_layer_3 = tf.layers.conv2d(inputs=maxpool_layer_2, filters=8,
kernel_size=(3,3), padding='same', activation=tf.nn.relu)

# The output tenosor will be in the shape of 4x4x8
encoded_layer = tf.layers.max_pooling2d(conv_layer_3, pool_size=(2,2),
strides=(2,2), padding='same')

# Defining the Decoder part of the netowrk
# Defining the first upsampling layer in the decoder part
# The output tenosor will be in the shape of 7x7x8
upsample_layer_1 = tf.image.resize_images(encoded_layer, size=(7,7),
method=tf.image.ResizeMethod.NEAREST_NEIGHBOR)

# The output tenosor will be in the shape of 7x7x8
conv_layer_4 = tf.layers.conv2d(inputs=upsample_layer_1, filters=8,
kernel_size=(3,3), padding='same', activation=tf.nn.relu)

# The output tenosor will be in the shape of 14x14x8
upsample_layer_2 = tf.image.resize_images(conv_layer_4, size=(14,14),
```

```
method=tf.image.ResizeMethod.NEAREST_NEIGHBOR)

# The output tenosor will be in the shape of 14x14x8
conv_layer_5 = tf.layers.conv2d(inputs=upsample_layer_2, filters=8,
kernel_size=(3,3), padding='same', activation=tf.nn.relu)

# The output tenosor will be in the shape of 28x28x8
upsample_layer_3 = tf.image.resize_images(conv_layer_5, size=(28,28),
method=tf.image.ResizeMethod.NEAREST_NEIGHBOR)

# The output tenosor will be in the shape of 28x28x16
conv6 = tf.layers.conv2d(inputs=upsample_layer_3, filters=16,
kernel_size=(3,3), padding='same', activation=tf.nn.relu)

# The output tenosor will be in the shape of 28x28x1
logits_layer = tf.layers.conv2d(inputs=conv6, filters=1, kernel_size=(3,3),
padding='same', activation=None)

# feeding the logits values to the sigmoid activation function to get the
reconstructed images
decoded_layer = tf.nn.sigmoid(logits_layer)

# feeding the logits to sigmoid while calculating the cross entropy
model_loss = tf.nn.sigmoid_cross_entropy_with_logits(labels=targets_values,
logits=logits_layer)

# Getting the model cost and defining the optimizer to minimize it
model_cost = tf.reduce_mean(model_loss)
model_optimizer =
tf.train.AdamOptimizer(learning_rate).minimize(model_cost)
```

Now we are good to go. We've built the decoder-decoder part of the convolutional neural network while showing how the dimensions of the input image will be reconstructed in the decoder part.

Model training

Now that we have the model built, we can kick off the learning process by generating random batches form the MNIST dataset and feed them to the optimizer defined earlier.

Let's start off by creating the session variable; it will be responsible for executing the computational graph that we defined earlier:

```
sess = tf.Session()
num_epochs = 20
```

```
train_batch_size = 200
sess.run(tf.global_variables_initializer())

for e in range(num_epochs):
    for ii in range(mnist_dataset.train.num_examples//train_batch_size):
        input_batch = mnist_dataset.train.next_batch(train_batch_size)
        input_images = input_batch[0].reshape((-1, 28, 28, 1))
        input_batch_cost, _ = sess.run([model_cost, model_optimizer],
feed_dict={inputs_values: input_images,targets_values: input_images})

        print("Epoch: {}/{}...".format(e+1, num_epochs),
              "Training loss: {:.3f}".format(input_batch_cost))
```

Output:

.

.

.

```
Epoch: 20/20... Training loss: 0.102
Epoch: 20/20... Training loss: 0.099
Epoch: 20/20... Training loss: 0.103
Epoch: 20/20... Training loss: 0.102
Epoch: 20/20... Training loss: 0.100
Epoch: 20/20... Training loss: 0.101
Epoch: 20/20... Training loss: 0.098
Epoch: 20/20... Training loss: 0.103
Epoch: 20/20... Training loss: 0.104
Epoch: 20/20... Training loss: 0.103
Epoch: 20/20... Training loss: 0.098
Epoch: 20/20... Training loss: 0.102
Epoch: 20/20... Training loss: 0.098
Epoch: 20/20... Training loss: 0.099
Epoch: 20/20... Training loss: 0.103
Epoch: 20/20... Training loss: 0.104
Epoch: 20/20... Training loss: 0.101
Epoch: 20/20... Training loss: 0.105
Epoch: 20/20... Training loss: 0.102
Epoch: 20/20... Training loss: 0.102
Epoch: 20/20... Training loss: 0.100
Epoch: 20/20... Training loss: 0.099
Epoch: 20/20... Training loss: 0.102
Epoch: 20/20... Training loss: 0.102
Epoch: 20/20... Training loss: 0.104
Epoch: 20/20... Training loss: 0.101
Epoch: 20/20... Training loss: 0.099
Epoch: 20/20... Training loss: 0.098
Epoch: 20/20... Training loss: 0.100
Epoch: 20/20... Training loss: 0.101
Epoch: 20/20... Training loss: 0.100
```

```
Epoch: 20/20... Training loss: 0.100
Epoch: 20/20... Training loss: 0.101
Epoch: 20/20... Training loss: 0.098
Epoch: 20/20... Training loss: 0.101
Epoch: 20/20... Training loss: 0.103
Epoch: 20/20... Training loss: 0.103
Epoch: 20/20... Training loss: 0.102
Epoch: 20/20... Training loss: 0.101
Epoch: 20/20... Training loss: 0.100
Epoch: 20/20... Training loss: 0.101
Epoch: 20/20... Training loss: 0.102
Epoch: 20/20... Training loss: 0.103
Epoch: 20/20... Training loss: 0.103
Epoch: 20/20... Training loss: 0.103
Epoch: 20/20... Training loss: 0.099
Epoch: 20/20... Training loss: 0.101
Epoch: 20/20... Training loss: 0.096
Epoch: 20/20... Training loss: 0.104
Epoch: 20/20... Training loss: 0.104
Epoch: 20/20... Training loss: 0.103
Epoch: 20/20... Training loss: 0.103
Epoch: 20/20... Training loss: 0.104
Epoch: 20/20... Training loss: 0.099
Epoch: 20/20... Training loss: 0.101
Epoch: 20/20... Training loss: 0.101
Epoch: 20/20... Training loss: 0.099
Epoch: 20/20... Training loss: 0.100
Epoch: 20/20... Training loss: 0.102
Epoch: 20/20... Training loss: 0.100
Epoch: 20/20... Training loss: 0.098
Epoch: 20/20... Training loss: 0.100
Epoch: 20/20... Training loss: 0.097
Epoch: 20/20... Training loss: 0.102
```

After running the preceding code snippet for 20 epochs, we'll get a trained CAE, so let's go ahead and test this model by feeding similar images from the MNIST dataset:

```
fig, axes = plt.subplots(nrows=2, ncols=10, sharex=True, sharey=True,
figsize=(20,4))
input_images = mnist_dataset.test.images[:10]
reconstructed_images = sess.run(decoded_layer, feed_dict={inputs_values:
input_images.reshape((10, 28, 28, 1))})

for imgs, row in zip([input_images, reconstructed_images], axes):
    for img, ax in zip(imgs, row):
        ax.imshow(img.reshape((28, 28)), cmap='Greys_r')
        ax.get_xaxis().set_visible(False)
        ax.get_yaxis().set_visible(False)
```

```
fig.tight_layout(pad=0.1)
```

Output:

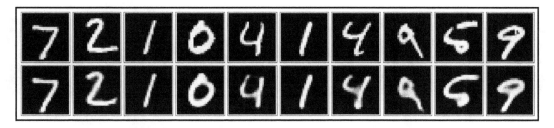

Figure 9: Examples of the original test images (first row) and their constructions (second row) using the convolution autoencoder

Denoising autoencoders

We can take the autoencoder architecture further by forcing it to learn more important features about the input data. By adding noise to the input images and having the original ones as the target, the model will try to remove this noise and learn important features about them in order to come up with meaningful reconstructed images in the output. This kind of CAE architecture can be used to remove noise from input images. This specific variation of autoencoders is called **denoising autoencoder**:

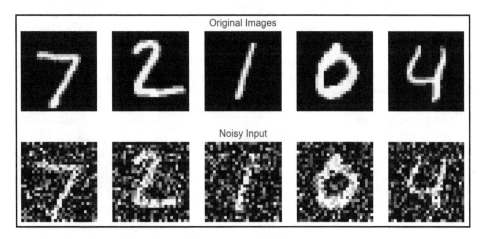

Figure 10: Examples of original images and the same images after adding a bit of Gaussian noise

So let's start off by implementing the architecture in the following figure. The only extra thing that we have added to this denoising autoencoder architecture is some noise in the original input image:

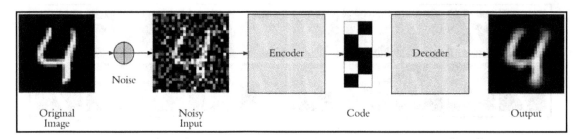

Figure 11: General denoising architecture of autoencoders

Building the model

In this implementation, we will be using more layers in the encoder and decoder part, and the reason for this is the new complexity that we have added to the input.

The next model is exactly the same as the previous CAE but with extra layers that will help us to reconstruct a noise-free image from a noisy one.

So let's go ahead and build this architecture:

```
learning_rate = 0.001

# Define the placeholder variable sfor the input and target values
inputs_values = tf.placeholder(tf.float32, (None, 28, 28, 1),
name='inputs_values')
targets_values = tf.placeholder(tf.float32, (None, 28, 28, 1),
name='targets_values')

# Defining the Encoder part of the netowrk
# Defining the first convolution layer in the encoder parrt
# The output tenosor will be in the shape of 28x28x32
conv_layer_1 = tf.layers.conv2d(inputs=inputs_values, filters=32,
kernel_size=(3,3), padding='same', activation=tf.nn.relu)

# The output tenosor will be in the shape of 14x14x32
maxpool_layer_1 = tf.layers.max_pooling2d(conv_layer_1, pool_size=(2,2),
strides=(2,2), padding='same')

# The output tenosor will be in the shape of 14x14x32
```

```
conv_layer_2 = tf.layers.conv2d(inputs=maxpool_layer_1, filters=32,
kernel_size=(3,3), padding='same', activation=tf.nn.relu)

# The output tenosor will be in the shape of 7x7x32
maxpool_layer_2 = tf.layers.max_pooling2d(conv_layer_2, pool_size=(2,2),
strides=(2,2), padding='same')

# The output tenosor will be in the shape of 7x7x16
conv_layer_3 = tf.layers.conv2d(inputs=maxpool_layer_2, filters=16,
kernel_size=(3,3), padding='same', activation=tf.nn.relu)

# The output tenosor will be in the shape of 4x4x16
encoding_layer = tf.layers.max_pooling2d(conv_layer_3, pool_size=(2,2),
strides=(2,2), padding='same')


# Defining the Decoder part of the netowrk
# Defining the first upsampling layer in the decoder part
# The output tenosor will be in the shape of 7x7x16
upsample_layer_1 = tf.image.resize_images(encoding_layer, size=(7,7),
method=tf.image.ResizeMethod.NEAREST_NEIGHBOR)

# The output tenosor will be in the shape of 7x7x16
conv_layer_4 = tf.layers.conv2d(inputs=upsample_layer_1, filters=16,
kernel_size=(3,3), padding='same', activation=tf.nn.relu)

# The output tenosor will be in the shape of 14x14x16
upsample_layer_2 = tf.image.resize_images(conv_layer_4, size=(14,14),
method=tf.image.ResizeMethod.NEAREST_NEIGHBOR)

# The output tenosor will be in the shape of 14x14x32
conv_layer_5 = tf.layers.conv2d(inputs=upsample_layer_2, filters=32,
kernel_size=(3,3), padding='same', activation=tf.nn.relu)

# The output tenosor will be in the shape of 28x28x32
upsample_layer_3 = tf.image.resize_images(conv_layer_5, size=(28,28),
method=tf.image.ResizeMethod.NEAREST_NEIGHBOR)

# The output tenosor will be in the shape of 28x28x32
conv_layer_6 = tf.layers.conv2d(inputs=upsample_layer_3, filters=32,
kernel_size=(3,3), padding='same', activation=tf.nn.relu)

# The output tenosor will be in the shape of 28x28x1
logits_layer = tf.layers.conv2d(inputs=conv_layer_6, filters=1,
kernel_size=(3,3), padding='same', activation=None)


# feeding the logits values to the sigmoid activation function to get the
```

```
reconstructed images
decoding_layer = tf.nn.sigmoid(logits_layer)

# feeding the logits to sigmoid while calculating the cross entropy
model_loss = tf.nn.sigmoid_cross_entropy_with_logits(labels=targets_values,
logits=logits_layer)

# Getting the model cost and defining the optimizer to minimize it
model_cost = tf.reduce_mean(model_loss)
model_optimizer =
tf.train.AdamOptimizer(learning_rate).minimize(model_cost)
```

Now we have a more complex or deeper version of the convolutional model.

Model training

It's time to start training this deeper network, which in turn will take more time to converge by reconstructing noise-free images from the noisy input.

So let's start off by creating the session variable:

```
sess = tf.Session()
```

Next up, we will kick off the training process but for more number of epochs:

```
num_epochs = 100
train_batch_size = 200

# Defining a noise factor to be added to MNIST dataset
mnist_noise_factor = 0.5
sess.run(tf.global_variables_initializer())

for e in range(num_epochs):
    for ii in range(mnist_dataset.train.num_examples//train_batch_size):
        input_batch = mnist_dataset.train.next_batch(train_batch_size)
        # Getting and reshape the images from the corresponding batch
        batch_images = input_batch[0].reshape((-1, 28, 28, 1))
        # Add random noise to the input images
        noisy_images = batch_images + mnist_noise_factor *
np.random.randn(*batch_images.shape)
        # Clipping all the values that are above 0 or above 1
        noisy_images = np.clip(noisy_images, 0., 1.)
        # Set the input images to be the noisy ones and the original images
to be the target
        input_batch_cost, _ = sess.run([model_cost, model_optimizer],
feed_dict={inputs_values: noisy_images,
```

```
                                                     targets_values:
batch_images}))

        print("Epoch: {}/{}...".format(e+1, num_epochs),
              "Training loss: {:.3f}".format(input_batch_cost))
```

Output:
.
.
.

```
Epoch: 100/100... Training loss: 0.098
Epoch: 100/100... Training loss: 0.101
Epoch: 100/100... Training loss: 0.103
Epoch: 100/100... Training loss: 0.098
Epoch: 100/100... Training loss: 0.102
Epoch: 100/100... Training loss: 0.102
Epoch: 100/100... Training loss: 0.103
Epoch: 100/100... Training loss: 0.101
Epoch: 100/100... Training loss: 0.098
Epoch: 100/100... Training loss: 0.099
Epoch: 100/100... Training loss: 0.096
Epoch: 100/100... Training loss: 0.100
Epoch: 100/100... Training loss: 0.100
Epoch: 100/100... Training loss: 0.103
Epoch: 100/100... Training loss: 0.100
Epoch: 100/100... Training loss: 0.101
Epoch: 100/100... Training loss: 0.099
Epoch: 100/100... Training loss: 0.096
Epoch: 100/100... Training loss: 0.102
Epoch: 100/100... Training loss: 0.099
Epoch: 100/100... Training loss: 0.098
Epoch: 100/100... Training loss: 0.102
Epoch: 100/100... Training loss: 0.100
Epoch: 100/100... Training loss: 0.100
Epoch: 100/100... Training loss: 0.099
Epoch: 100/100... Training loss: 0.098
Epoch: 100/100... Training loss: 0.100
Epoch: 100/100... Training loss: 0.099
Epoch: 100/100... Training loss: 0.102
Epoch: 100/100... Training loss: 0.099
Epoch: 100/100... Training loss: 0.102
Epoch: 100/100... Training loss: 0.100
Epoch: 100/100... Training loss: 0.101
Epoch: 100/100... Training loss: 0.102
Epoch: 100/100... Training loss: 0.098
Epoch: 100/100... Training loss: 0.103
Epoch: 100/100... Training loss: 0.100
Epoch: 100/100... Training loss: 0.098
```

```
Epoch: 100/100... Training loss: 0.100
Epoch: 100/100... Training loss: 0.097
Epoch: 100/100... Training loss: 0.099
Epoch: 100/100... Training loss: 0.100
Epoch: 100/100... Training loss: 0.101
Epoch: 100/100... Training loss: 0.101
```

Now we have trained the model to be able to produce noise-free images, which makes autoencoders applicable to many domains.

In the next snippet of code, we will not feed the row images of the MNIST test set to the model as we need to add noise to these images first to see how the trained model will be able to produce noise-free images.

Here I'm adding noise to the test images and passing them through the autoencoder. It does a surprisingly great job of removing the noise, even though it's sometimes difficult to tell what the original number is:

```
#Defining some figures
fig, axes = plt.subplots(nrows=2, ncols=10, sharex=True, sharey=True,
figsize=(20,4))

#Visualizing some images
input_images = mnist_dataset.test.images[:10]
noisy_imgs = input_images + mnist_noise_factor *
np.random.randn(*input_images.shape)

#Clipping and reshaping the noisy images
noisy_images = np.clip(noisy_images, 0., 1.).reshape((10, 28, 28, 1))

#Getting the reconstructed images
reconstructed_images = sess.run(decoding_layer, feed_dict={inputs_values:
noisy_images})

#Visualizing the input images and the noisy ones
for imgs, row in zip([noisy_images, reconstructed_images], axes):
    for img, ax in zip(imgs, row):
        ax.imshow(img.reshape((28, 28)), cmap='Greys_r')
        ax.get_xaxis().set_visible(False)
        ax.get_yaxis().set_visible(False)

fig.tight_layout(pad=0.1)
```

Output:

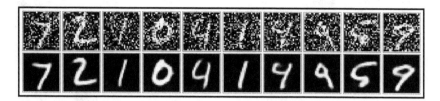

Figure 12: Examples of original test images with some Gaussian noise (top row) and their construction based on the trained denoising autoencoder

Applications of autoencoders

In the previous example of constructing images from a lower representation, we saw it was very similar to the original input, and also we saw the benefits of CANs while denoising the noisy dataset. This kind of example we have implemented above is really useful for the image construction applications and dataset denoising. So you can generalize the above implementation to any other example of interest to you.

Also, throughout this chapter, we have seen how flexible the autoencoder architecture is and how we can make different changes to it. We have even tested it to solve harder problems of removing noise from input images. This kind of flexibility opens the door to many more applications that auoencoders will be a great fit for.

Image colorization

Autoencoders—especially the convolutional version—can be used for harder tasks such as image colorization. In the following example, we feed the model with an input image without any colors, and the reconstructed version of this image will be colorized by the autoencoder model:

Figure 13: The CAE is trained to colorize the image

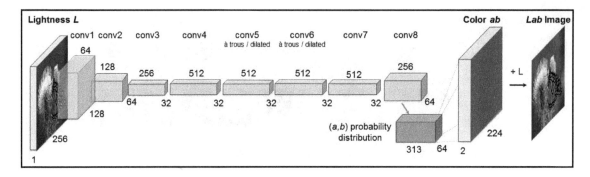

Figure 14: Colorization paper architecture

Now that our autoencoder is trained, we can use it to colorize pictures we have never seen before!

This kind of application can be used to color very old images that were taken in the early days of the camera.

More applications

Another interesting application can be producing images with higher resolution, or neural image enhancement, like the following figures show.

These figures show more realistic versions of image colorization by Richard Zhang:

Figure 15: Colorful image colorization by Richard Zhang, Phillip Isola, and Alexei A. Efros

This figure shows another application of autoencoders to make image enhancements:

Figure 16: Neural enhancement by Alexjc (https://github.com/alexjc/neural-enhance)

Summary

In this chapter, we introduced a totally new architecture that can be used for many interesting applications. Autoencoders are very flexible, so feel free to come up with your own problem in the area of image enhancement, colorization, or construction. Also, there are more variations of autoencoders, called **variational autoencoders**. They are also used for very interesting applications, such as image generation.

14
Generative Adversarial Networks

Generative Adversarial Networks (**GANs**) are deep neural net architectures that consist of two networks pitted against each other (hence the name **adversarial**).

GANs were introduced in a paper (https://arxiv.org/abs/1406.2661) by Ian Goodfellow and other researchers, including Yoshua Bengio, at the University of Montreal in 2014. Referring to GANs, Facebook's AI research director, Yann LeCun, called **adversarial training** the most interesting idea in the last 10 years in machine learning.

The potential of GANs is huge, because they can learn to mimic any distribution of data. That is, GANs can be taught to create worlds eerily similar to our own in any domain: images, music, speech, or prose. They are robot artists in a sense, and their output is impressive (https://www.nytimes.com/2017/08/14/arts/design/google-how-ai-creates-new-music-and-new-artists-project-magenta.html)—and poignant too.

The following topics will be covered in this chapter:

- An intuitive introduction
- Simple implementation of GANs
- Deep convolutional GANs

An intuitive introduction

In this section, we are going to introduce GANs in a very intuitive way. To get an idea of how GANs work, we will adopt a fake scenario of getting a ticket for a party.

The story starts with a very interesting party or event being held somewhere, and you are very interested in attending it. You hear about this event very late and all the tickets are sold out, but you will do anything to get into the party. So you come up with an idea! You will try to fake a ticket that needs to be exactly the same as the original one, or very, very similar to it. But because life is not easy, there's another challenge: you don't know what the original ticket looks like. So from your experience of going to such parties, you start to imagine what the ticket might look like and start to design the ticket based on your imagination.

You will try to design the ticket and then go to the event and show the ticket to the security guys. Hopefully, they will be convinced and will let you in. But you don't want to show your face multiple times to the security guards, so you decide to get help from your friend, who will take your initial guess about the original ticket and show it to the security guards. If they don't let him in, he will get some information for you about what the ticket might look like, based on seeing some people getting in with the actual ticket. You will refine the ticket based on your friend's comments until the security guards let him in. At this point—and at this point only—you will design another one that has exactly the same look and get yourself in.

Do, think too much about how unrealistic this story is, but the way GANs work is very similar to this story. GANs are very trendy nowadays, and people are using them for many applications in the field of computer vision.

There are many interesting applications that you can use GANs for, and we will implement and mention some of them.

In GANs, there are two main components that have made a breakthrough in many computer vision fields. The first component is called **Generator** and the second one is called **Discriminator**:

- The Generator will try to generate data samples out of a specific probability distribution, which is very similar to the guy who was trying to replicate a ticket for the event

- The Discriminator will judge (like the security guys who are trying to find flaws in the ticket to decide whether it's original or fake) whether its input is coming from the original training set (an original ticket) or from the generator part (designed by the guy who's trying to replicate the original ticket):

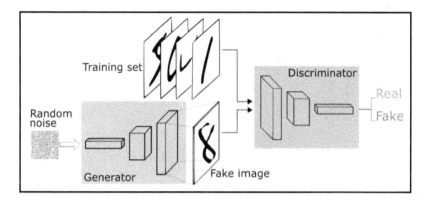

Figure 1: GANs – general architecture

Simple implementation of GANs

From the story of faking a ticket to an event, the idea of GANs seems to be very intuitive. So to get a clear understanding of how GANs work and how to implement them, we are going to demonstrate a simple implementation of a GAN on the MNIST dataset.

First, we need to build the core of the GAN network, which is comprised of two major components: the generator and the discriminator. As we said, the generator will try to imagine or fake data samples from a specific probability distribution; the discriminator, which has access to and sees the actual data samples, will judge whether the generator's output has any flaws in the design or it's very close to the original data samples. Similar to the scenario of the event, the whole purpose of the generator is to try to convince the discriminator that the generated image is from the real dataset and hence try to fool him.

The training process has a similar end to the event story; the generator will finally manage to generate images that look very similar to the original data samples:

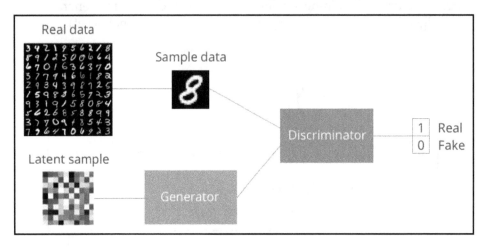

Figure 2: GAN general architecture for the MNIST dataset

The typical structure of any GAN is shown in *Figure 2*, which will be trained on the MNIST dataset. The `Latent sample` part in this figure is a random thought or vector that the generator uses to replicate the real images with fake ones.

As we mentioned, the discriminator works as a judge and it will try to separate the real images from the fake ones that were designed by the generator. So the output of this network will be binary, which can be represented by a sigmoid function with 0 (meaning the input is a fake image) and 1 (meaning that the input is a real image).

Let's go ahead and start implementing this architecture to see how it performs on the MNIST dataset.

Let's start of by importing the required libraries for this implementation:

```
%matplotlib inline

import matplotlib.pyplot as plt
import pickle as pkl

import numpy as np
import tensorflow as tf
```

We will be using the MNIST dataset, so we are going to use TensorFlow helpers to get the dataset and store it somewhere:

```
from tensorflow.examples.tutorials.mnist import input_data
mnist_dataset = input_data.read_data_sets('MNIST_data')

Output:
Extracting MNIST_data/train-images-idx3-ubyte.gz
Extracting MNIST_data/train-labels-idx1-ubyte.gz
Extracting MNIST_data/t10k-images-idx3-ubyte.gz
Extracting MNIST_data/t10k-labels-idx1-ubyte.gz
```

Model inputs

Before diving into building the core of the GAN, which is represented by the generator and discriminator, we are going to define the inputs of our computational graph. As shown in *Figure 2*, we need two inputs. The first one will be the real images, which will be fed to the discriminator. The other input is called **latent space**, which will be fed to the generator and used to generate its fake images:

```
# Defining the model input for the generator and discrimator
def inputs_placeholders(discrimator_real_dim, gen_z_dim):
    real_discrminator_input = tf.placeholder(tf.float32, (None,
discrimator_real_dim), name="real_discrminator_input")
    generator_inputs_z = tf.placeholder(tf.float32, (None, gen_z_dim),
name="generator_input_z")
    return real_discrminator_input, generator_inputs_z
```

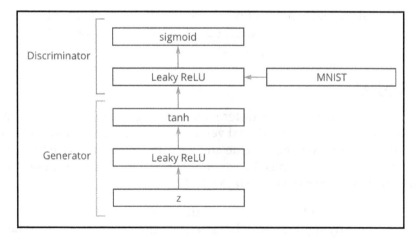

Figure 3: Architecture of the MNIST GAN implementation

Now it's time to dive into building the two core components of our architecture. We will start by building the generator part. As shown in *Figure 3*, the generator will consist of at least one hidden layer, which will work as an approximator. Also, instead of using the normal ReLU activation function, we will use something called a leaky ReLU. This will allow the gradient values to flow through the layer without any constraints (more in the next section about leaky RelU).

Variable scope

Variable scope is a feature of TensorFlow that helps us to do the following:

- Make sure that we have some naming conventions to retrieve them later, for example, by making them start with the word generator or discriminator, which will help us during the training of the network. We could have used the name scope feature, but this feature won't help us for the second purpose.
- To be able to reuse or retrain the same network but with different inputs. For example, we are going to sample fake images from the generator to see how good the generator is for replicating the original ones. Also, the discriminator will have access to the real and fake images, which will make it easy for us to reuse the variables instead of creating new ones while building the computational graph.

The following statement will show how to use the variable scope feature of TensorFlow:

```
with tf.variable_scope('scopeName', reuse=False):
    # Write your code here
```

You can read more about the benefits of using the variable scope feature at `https://www.tensorflow.org/programmers_guide/variable_scope#the_problem`.

Leaky ReLU

We mentioned that we will be using a different version than the ReLU activation function, which is called leaky ReLU. The traditional version of the ReLU activation function will just take the maximum between the input value and zero, by other means truncating negative values to zero. Leaky ReLU, which is the version that we will be using, allows some negative values to exist, hence the name **leaky ReLU**.

Sometimes, if we use the traditional ReLU activation function, the network gets stuck in a popular state called the dying state, and that's because the network produces nothing but zeros for all the outputs.

The idea of using leaky ReLU is to prevent this dying state by allowing some negative values to pass through.

The whole idea behind making the generator work is to receive gradient values from the discriminator, and if the network is stuck in a dying situation, the learning process won't happen.

The following figures illustrate the difference between traditional ReLU and its leaky version:

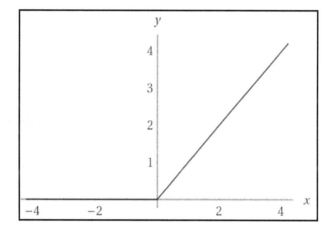

Figure 4: ReLU function

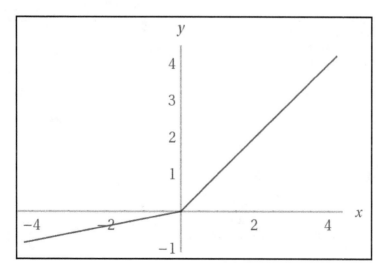

Figure 5: Leaky ReLU activation functions

The leaky ReLU activation function is not implemented in TensorFlow, so we need to implement it ourselves. The output of this activation function will be positive if the input is positive, and will be a controlled negative value if the input is negative. We will control the negative value by a parameter called **alpha**, which will introduce tolerance of the network by allowing some negative values to pass.

The following equation represents the leaky ReLU that we will be implementing:

$$f(x) = max(a \times x, x)$$

Generator

The MNIST images are normalized between 0 and 1, where the sigmoid activation function can work best. But in practice, the tanh activation function is found to give better performance than any other function. So, in order to use the tanh activation function, we will need to re-scale the range of the pixel values of these images to be between -1 and 1:

```
def generator(gen_z, gen_out_dim, num_hiddern_units=128, reuse_vars=False,
leaky_relu_alpha=0.01):
    ''' Building the generator part of the network
        Function arguments
        ---------
        gen_z : the generator input tensor
        gen_out_dim : the output shape of the generator
        num_hiddern_units : Number of neurons/units in the hidden layer
        reuse_vars : Reuse variables with tf.variable_scope
        leaky_relu_alpha : leaky ReLU parameter
        Function Returns
        -------
        tanh_output, logits_layer:
    '''
    with tf.variable_scope('generator', reuse=reuse_vars):
        # Defining the generator hidden layer
        hidden_layer_1 = tf.layers.dense(gen_z, num_hiddern_units,
activation=None)
        # Feeding the output of hidden_layer_1 to leaky relu
        hidden_layer_1 = tf.maximum(hidden_layer_1,
leaky_relu_alpha*hidden_layer_1)
        # Getting the logits and tanh layer output
        logits_layer = tf.layers.dense(hidden_layer_1, gen_out_dim,
activation=None)
        tanh_output = tf.nn.tanh(logits_layer)
        return tanh_output, logits_layer
```

Now we have the generator part ready. Let's go ahead and define the second component of the network.

Discriminator

Next up, we will build the second main component in the generative adversarial network, which is the discriminator. The discriminator is pretty much the same as the generator, but instead of using the `tanh` activation function, we will be using the `sigmoid` activation function; it will produce a binary output that will represent the judgment of the discriminator on the input image:

```
def discriminator(disc_input, num_hiddern_units=128, reuse_vars=False,
leaky_relu_alpha=0.01):
    ''' Building the discriminator part of the network
        Function Arguments
        ---------
        disc_input : discrminator input tensor
        num_hiddern_units : Number of neurons/units in the hidden layer
        reuse_vars : Reuse variables with tf.variable_scope
        leaky_relu_alpha : leaky ReLU parameter
        Function Returns
        -------
        sigmoid_out, logits_layer:
    '''
    with tf.variable_scope('discriminator', reuse=reuse_vars):
        # Defining the generator hidden layer
        hidden_layer_1 = tf.layers.dense(disc_input, num_hiddern_units,
activation=None)
        # Feeding the output of hidden_layer_1 to leaky relu
        hidden_layer_1 = tf.maximum(hidden_layer_1,
leaky_relu_alpha*hidden_layer_1)
        logits_layer = tf.layers.dense(hidden_layer_1, 1, activation=None)
        sigmoid_out = tf.nn.sigmoid(logits_layer)
        return sigmoid_out, logits_layer
```

Building the GAN network

After defining the main functions that will build the generator and discriminator parts, it's time to stack them up and define the model losses and optimizers for this implementation.

Model hyperparameters

We can fine-tune the GANs by changing the following set of hyperparameters:

```
# size of discriminator input image
#28 by 28 will flattened to be 784
input_img_size = 784

# size of the generator latent vector
gen_z_size = 100

# number of hidden units for the generator and discriminator hidden layers
gen_hidden_size = 128
disc_hidden_size = 128

#leaky ReLU alpha parameter which controls the leak of the function
leaky_relu_alpha = 0.01

# smoothness of the label
label_smooth = 0.1
```

Defining the generator and discriminator

After defining the two main parts of our architecture that will be used to generate fake MNIST images (which will look exactly the same as the real ones), it's time to build the network using the functions that we have defined so far. In order to build the network, we are going to follow these steps:

1. Defining the inputs for our model, which will consist of two variables. One of these variables will be the real images, which will be fed to the discriminator, and the second will be the latent space to be used by the generator to replicate the original images.
2. Calling the defined generator function to build the generator part of the network.
3. Calling the defined discriminator function to build the discriminator part of the network, but we are going to call this function twice. One call will be for the real data and the second call will be for the fake data from the generator.
4. Keeping the weights of real and fake images the same by reusing the variables:

   ```
   tf.reset_default_graph()

   # creating the input placeholders for the discrminator and
   generator
   real_discrminator_input, generator_input_z =
   inputs_placeholders(input_img_size, gen_z_size)
   ```

```
#Create the generator network
gen_model, gen_logits = generator(generator_input_z,
input_img_size, gen_hidden_size, reuse_vars=False,
leaky_relu_alpha=leaky_relu_alpha)

# gen_model is the output of the generator
#Create the generator network
disc_model_real, disc_logits_real =
discriminator(real_discrminator_input, disc_hidden_size,
reuse_vars=False, leaky_relu_alpha=leaky_relu_alpha)
disc_model_fake, disc_logits_fake = discriminator(gen_model,
disc_hidden_size, reuse_vars=True,
leaky_relu_alpha=leaky_relu_alpha)
```

Discriminator and generator losses

In this part, we need to define the discriminator and generator losses, and this can be considered to be the most tricky part of this implementation.

We know that the generator tries to replicate the original images and that the discriminator works as a judge, receiving both images from the generator and the original input images. So while designing our loss for each part, we need to target two things.

First, we need the discriminator part of the network to be able to distinguish between the fake images generated by the generator and the real images coming from the original training examples. During training time, we will feed the discriminator part with a batch that is divided into two categories. The first category will be images from the original input and the second category will be images from the fake ones that got generated by the generator.

So the final general loss of the discriminator will be the sum of its ability to accept the real ones as real and detect the fake ones as fake; then the final total loss will be:

$$\text{disc\_loss} = \text{disc\_loss\_real} + \text{disc\_loss\_fake}$$

```
tf.reduce_mean(tf.nn.sigmoid_cross_entropy_with_logits(logits=logits_layer,
labels=labels))
```

So we need to calculate two losses to come up with the final discriminator loss.

The first loss, `disc_loss_real`, will be calculated based on the `logits` values that we will get from the discriminator and the `labels`, which will be all ones in this case since we know that all the images in this mini-batch are all coming from the real input images of the MNIST dataset. To enhance the ability of the model to generalize on the test set and give better results, people have found that practically changing the value of 1 to 0.9 is better. This kind of change to the label introduces something called **label smooth**:

```
labels = tf.ones_like(tensor) * (1 - smooth)
```

For the second part of the discriminator loss, which is the ability of the discriminator to detect fake images, the loss will be between the logits values that we will get from the discriminator and labels; all of these are zeros since we know that all the images in this mini-batch are coming from the generator, and not from the original input.

Now that we have discussed the discriminator loss, we need to calculate the generator loss as well. The generator loss will be called `gen_loss`, which will be the loss between `disc_logits_fake` (the output of the discriminator for the fake images) and the labels (which will be all ones since the generator is trying to convince the discriminator with its design of the fake image):

```
# calculating the losses of the discrimnator and generator
disc_labels_real = tf.ones_like(disc_logits_real) * (1 - label_smooth)
disc_labels_fake = tf.zeros_like(disc_logits_fake)

disc_loss_real =
tf.nn.sigmoid_cross_entropy_with_logits(labels=disc_labels_real,
logits=disc_logits_real)
disc_loss_fake =
tf.nn.sigmoid_cross_entropy_with_logits(labels=disc_labels_fake,
logits=disc_logits_fake)

#averaging the disc loss
disc_loss = tf.reduce_mean(disc_loss_real + disc_loss_fake)

#averaging the gen loss
gen_loss = tf.reduce_mean(
    tf.nn.sigmoid_cross_entropy_with_logits(
        labels=tf.ones_like(disc_logits_fake),
        logits=disc_logits_fake))
```

Optimizers

Finally, the optimizers part! In this section, we will define the optimization criteria that will be used during the training process. First off, we are going to update the variables of the generator and discriminator separately, so we need to be able to retrieve the variables of each part.

For the first optimizer, the generator one, we will retrieve all the variables that start with the name `generator` from the trainable variables of the computational graph; then we can check which variable is which by referring to its name.

We'll do the same for the discriminator variables as well, by letting in all variables that start with `discriminator`. After that, we can pass the list of variables that we want to be optimized to the optimizer.

So the variable scope feature of TensorFlow gave us the ability to retrieve variables that start with a certain string, and then we can have two different lists of variables, one for the generator and another one for the discriminator:

```
# building the model optimizer

learning_rate = 0.002

# Getting the trainable_variables of the computational graph, split into
Generator and Discrimnator parts
trainable_vars = tf.trainable_variables()
gen_vars = [var for var in trainable_vars if
var.name.startswith("generator")]
disc_vars = [var for var in trainable_vars if
var.name.startswith("discriminator")]

disc_train_optimizer = tf.train.AdamOptimizer().minimize(disc_loss,
var_list=disc_vars)
gen_train_optimizer = tf.train.AdamOptimizer().minimize(gen_loss,
var_list=gen_vars)
```

Model training

Now let's kick off the training process and see how GANs will manage to generate images similar to the MNIST ones:

```
train_batch_size = 100
num_epochs = 100
generated_samples = []
```

```
model_losses = []

saver = tf.train.Saver(var_list = gen_vars)

with tf.Session() as sess:
    sess.run(tf.global_variables_initializer())
    for e in range(num_epochs):
        for ii in
range(mnist_dataset.train.num_examples//train_batch_size):
            input_batch = mnist_dataset.train.next_batch(train_batch_size)
            # Get images, reshape and rescale to pass to D
            input_batch_images = input_batch[0].reshape((train_batch_size,
784))
            input_batch_images = input_batch_images*2 - 1
            # Sample random noise for G
            gen_batch_z = np.random.uniform(-1, 1, size=(train_batch_size,
gen_z_size))
            # Run optimizers
            _ = sess.run(disc_train_optimizer,
feed_dict={real_discrminator_input: input_batch_images, generator_input_z:
gen_batch_z})
            _ = sess.run(gen_train_optimizer, feed_dict={generator_input_z:
gen_batch_z})
        # At the end of each epoch, get the losses and print them out
        train_loss_disc = sess.run(disc_loss, {generator_input_z:
gen_batch_z, real_discrminator_input: input_batch_images})
        train_loss_gen = gen_loss.eval({generator_input_z: gen_batch_z})
        print("Epoch {}/{}...".format(e+1, num_epochs),
            "Disc Loss: {:.3f}...".format(train_loss_disc),
            "Gen Loss: {:.3f}".format(train_loss_gen))
        # Save losses to view after training
        model_losses.append((train_loss_disc, train_loss_gen))
        # Sample from generator as we're training for
viegenerator_inputs_zwing afterwards
        gen_sample_z = np.random.uniform(-1, 1, size=(16, gen_z_size))
        generator_samples = sess.run(
                    generator(generator_input_z, input_img_size,
reuse_vars=True),
                    feed_dict={generator_input_z: gen_sample_z})
        generated_samples.append(generator_samples)
        saver.save(sess, './checkpoints/generator_ck.ckpt')

# Save training generator samples
with open('train_generator_samples.pkl', 'wb') as f:
    pkl.dump(generated_samples, f)

Output:
    .
```

.
.

```
Epoch 71/100... Disc Loss: 1.078... Gen Loss: 1.361
Epoch 72/100... Disc Loss: 1.037... Gen Loss: 1.555
Epoch 73/100... Disc Loss: 1.194... Gen Loss: 1.297
Epoch 74/100... Disc Loss: 1.120... Gen Loss: 1.730
Epoch 75/100... Disc Loss: 1.184... Gen Loss: 1.425
Epoch 76/100... Disc Loss: 1.054... Gen Loss: 1.534
Epoch 77/100... Disc Loss: 1.457... Gen Loss: 0.971
Epoch 78/100... Disc Loss: 0.973... Gen Loss: 1.688
Epoch 79/100... Disc Loss: 1.324... Gen Loss: 1.370
Epoch 80/100... Disc Loss: 1.178... Gen Loss: 1.710
Epoch 81/100... Disc Loss: 1.070... Gen Loss: 1.649
Epoch 82/100... Disc Loss: 1.070... Gen Loss: 1.530
Epoch 83/100... Disc Loss: 1.117... Gen Loss: 1.705
Epoch 84/100... Disc Loss: 1.042... Gen Loss: 2.210
Epoch 85/100... Disc Loss: 1.152... Gen Loss: 1.260
Epoch 86/100... Disc Loss: 1.327... Gen Loss: 1.312
Epoch 87/100... Disc Loss: 1.069... Gen Loss: 1.759
Epoch 88/100... Disc Loss: 1.001... Gen Loss: 1.400
Epoch 89/100... Disc Loss: 1.215... Gen Loss: 1.448
Epoch 90/100... Disc Loss: 1.108... Gen Loss: 1.342
Epoch 91/100... Disc Loss: 1.227... Gen Loss: 1.468
Epoch 92/100... Disc Loss: 1.190... Gen Loss: 1.328
Epoch 93/100... Disc Loss: 0.869... Gen Loss: 1.857
Epoch 94/100... Disc Loss: 0.946... Gen Loss: 1.740
Epoch 95/100... Disc Loss: 0.925... Gen Loss: 1.708
Epoch 96/100... Disc Loss: 1.067... Gen Loss: 1.427
Epoch 97/100... Disc Loss: 1.099... Gen Loss: 1.573
Epoch 98/100... Disc Loss: 0.972... Gen Loss: 1.884
Epoch 99/100... Disc Loss: 1.292... Gen Loss: 1.610
Epoch 100/100... Disc Loss: 1.103... Gen Loss: 1.736
```

After running the model for 100 epochs, we have a trained model that will be able to generate images similar to the original input images that we fed to the discriminator:

```
fig, ax = plt.subplots()
model_losses = np.array(model_losses)
plt.plot(model_losses.T[0], label='Disc loss')
plt.plot(model_losses.T[1], label='Gen loss')
plt.title("Model Losses")
plt.legend()
```

Output:

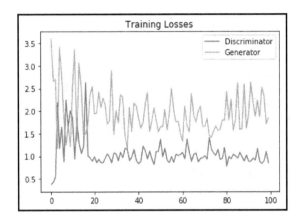

Figure 6: Discriminator and Generator Losses

As shown in the preceding figure, you can see that the model losses, which are represented by the Discriminator and Generator lines, are converging.

Generator samples from training

Let's test the performance of the model and even see how the generation skills (designing tickets for the event) of the generator got enhanced while approaching the end of the training process:

```
def view_generated_samples(epoch_num, g_samples):
    fig, axes = plt.subplots(figsize=(7,7), nrows=4, ncols=4, sharey=True,
sharex=True)
    print(gen_samples[epoch_num][1].shape)
    for ax, gen_image in zip(axes.flatten(), g_samples[0][epoch_num]):
        ax.xaxis.set_visible(False)
        ax.yaxis.set_visible(False)
        img = ax.imshow(gen_image.reshape((28,28)), cmap='Greys_r')
    return fig, axes
```

Before plotting some generated images from the last epoch in the training process, we need to load the persisted file that contains generated samples at each epoch during the training process:

```
# Load samples from generator taken while training
with open('train_generator_samples.pkl', 'rb') as f:
    gen_samples = pkl.load(f)
```

Now, let's plot the 16 generated images from the last epoch of the training process and see how the generator was able to generate meaningful numbers such as 3, 7, and 2:

```
_ = view_generated_samples(-1, gen_samples)
```

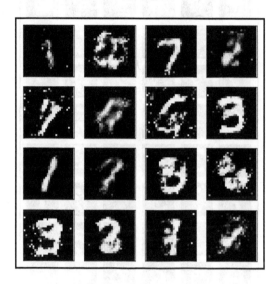

Figure 7: Samples from the final training epoch

We can even see the designing skills of the generator over different epochs. So let's visualize the images that are generated by it in every 10 epochs:

```
rows, cols = 10, 6
fig, axes = plt.subplots(figsize=(7,12), nrows=rows, ncols=cols,
sharex=True, sharey=True)

for gen_sample, ax_row in zip(gen_samples[::int(len(gen_samples)/rows)],
axes):
    for image, ax in zip(gen_sample[::int(len(gen_sample)/cols)], ax_row):
        ax.imshow(image.reshape((28,28)), cmap='Greys_r')
        ax.xaxis.set_visible(False)
        ax.yaxis.set_visible(False)
```

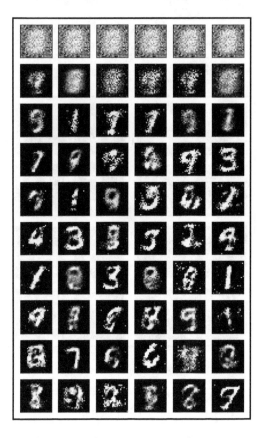

Figure 8: Images generated as the network was training, for every 10 epochs

As you can see, the designing skills of the generator and its ability to generate fake images were very limited at first, and then it was enhanced towards the end of the training process.

Sampling from the generator

In the previous section, we went through some examples that were generated during the training process of this GAN architecture. We can also generate completely new images from the generator by loading the checkpoints that we have saved and feeding the generator with a new latent space that it can use to generate new images:

```
# Sampling from the generator
saver = tf.train.Saver(var_list=g_vars)

with tf.Session() as sess:
```

```
#restoring the saved checkpints
saver.restore(sess, tf.train.latest_checkpoint('checkpoints'))
gen_sample_z = np.random.uniform(-1, 1, size=(16, z_size))
generated_samples = sess.run(
            generator(generator_input_z, input_img_size,
reuse_vars=True),
            feed_dict={generator_input_z: gen_sample_z})
view_generated_samples(0, [generated_samples])
```

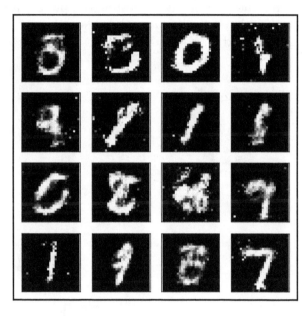

Figure 9: Samples from the generator

There are some observations that you can come up with while implementing this example. During the first epochs of the training process, the generator doesn't have any skills to produce similar images to the real one because it doesn't know what they look like. Even the discriminator doesn't know how to distinguish between fake images made by the generator and the. At the beginning of training, two interesting situations occur. First, the generator does not know how to create images like the real ones that we fed originally to the network. Second, the discriminator doesn't know the difference between the real and fake images.

Later on, the generator starts to fake images that make sense to some extent, and that's because the generator will learn the data distribution that the original input images are coming from. In parallel, the discriminator will be able to distinguish between fake and real images and it will be fooled by the end of the training process.

Summary

GANs are being used nowadays for lots of interesting applications. GANs can be used for different setups such as semi-supervised and unsupervised tasks. Also, because of the huge number of researchers working on GANs, these models are progressing day by day and their ability to generate images or videos is getting better and better.

These kinds of models can be used for many interesting commercial applications, such as adding a plugin to Photoshop that can take commands like `make my smile more appealing`. They can also be used for image denoising.

15
Face Generation and Handling Missing Labels

The list of interesting applications that we can use GANs for is endless. In this chapter, we are going to demonstrate another promising application of GANs, which is face generation based on the CelebA database. We'll also demonstrate how to use GANs for semi-supervised learning setups where we've got a poorly labeled dataset with some missing labels.

The following topics will be covered in this chapter:

- Face generation
- Semi-supervised learning with generative adversarial networks (GANs)

Face generation

As we mentioned in the previous chapter, the Generator and Discriminator consist of a **Deconvolutional Network** (**DNN**: https://www.quora.com/How-does-a-deconvolutional-neural-network-work) and **Convolutional Neural Network** (**CNN**: http://cs231n.github.io/convolutional-networks/):

- CNN is a a type of neural network that encodes hundreds of pixels of an image into a vector of small dimensions (z), which is a summary of the image
- DNN is a network that learns some filters to recover the original image from z

Also, the discriminator will output one or zero to indicate whether the input image is from the actual dataset or generated by the generator. On the other side, the generator will try to replicate images similar to the original dataset based on the latent space z, which might follow a Gaussian distribution. So, the goal of the discriminator is to correctly discriminate between the real images, and the goal of the generator is to learn the distribution of the original dataset and hence fool the discriminator so that it makes a wrong decision.

In this section, we'll try to teach the generator to learn human face image distribution so that it can generate realistic faces.

Generating human-like faces is crucial for most graphics companies, who are always looking for new faces for their applications, and it gives us a clue of how artificial intelligence is really close to achieving realism in generating artificial faces.

In this example, we'll be using the CelebA dataset. The CelebFaces Attributes Dataset (CelebA) is a large-scale face attributes dataset with about 200K celebrity images, each with 40 attribute annotations. There are lots of pose variations covered by the dataset, as well as background clutter, so CelebA is very diverse and well annotated. it includes:

- 10,177 identities
- 202,599 face images
- Five landmark locations and 40 binary attribute annotations per image

We can use this dataset for many computer vision applications other than face generation, such as face recognition and localization, or face attribute detection.

This figure shows how the generator error, or learning human face distribution, gets close to realism during the training process:

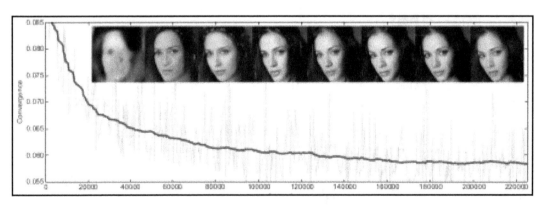

Figure 1: GANs for generating new faces from a celebrity images dataset

Getting the data

In this section, we will define some helper functions that will help us to download the CelebA dataset. We'll start off by importing their required packages for this implementation:

```
import math
import os
import hashlib
from urllib.request import urlretrieve
import zipfile
import gzip
import shutil

import numpy as np
from PIL import Image
from tqdm import tqdm
import utils

import tensorflow as tf
```

Next up, we are going to use the utils script to download the dataset:

```
#Downloading celebA dataset
celebA_data_dir = 'input'
utils.download_extract('celeba', celebA_data_dir)

Output:

Downloading celeba: 1.44GB [00:21, 66.6MB/s]
Extracting celeba...
```

Exploring the Data

The CelebA dataset contains over 200k annotated celebrity images. Since we are going to use GANs to generate similar images, it is worth looking at a bunch of images from the dataset and see how they look. In this section, we are going to define some helper functions for visualizing a bunch of images from the CelebA dataset.

Now, let's use the `utils` script to display some images from the dataset:

```
#number of images to display
num_images_to_show = 25

celebA_images = utils.get_batch(glob(os.path.join(celebA_data_dir,
```

```
'img_align_celeba/*.jpg'))[:num_images_to_show], 28,
                        28, 'RGB')
pyplot.imshow(utils.images_square_grid(celebA_images, 'RGB'))
```

Output:

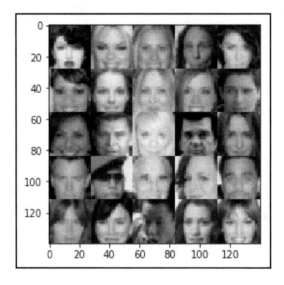

Figure 2: Plotting a sample of images from CelebA dataset

The main focus of this computer vision task is to use GANs for generating images similar two the ones in the celebrity dataset, so we'll need to focus on the face part of the images. To focus on the face part of an image, we are going to remove the parts of the image that don't include a celebrity face.

Building the model

Now, let's start off by building the core of our implementation, which is the computational graph; it will mainly include the following components:

- Model inputs
- Discriminator
- Generator
- Model losses
- Model optimizer
- Training the model

Model inputs

In this section, we are going to implement a helper function that we'll define the model input placeholders which will responsible for feeding the data the the computational graph.

The functions should be able to create three main placeholders:

- Actual input images from the dataset which will have the dimensions of (batch size, input image width, input image height, number of channels)
- The latent space Z, which will be used by the generator for generating fake images
- Learning rate placeholder

The helper function will return a tuple of these three input placeholders. So, let's go ahead and define this function:

```
# defining the model inputs
def inputs(img_width, img_height, img_channels, latent_space_z_dim):
    true_inputs = tf.placeholder(tf.float32, (None, img_width, img_height,
img_channels),
                                 'true_inputs')
    l_space_inputs = tf.placeholder(tf.float32, (None, latent_space_z_dim),
'l_space_inputs')
    model_learning_rate = tf.placeholder(tf.float32,
name='model_learning_rate')

    return true_inputs, l_space_inputs, model_learning_rate
```

Discriminator

Next up, we need to implement the discriminator part of the network, which will be used to judge whether the incoming input is coming from the real dataset or generated by the generator. Again, we'll use the TensorFlow feature of `tf.variable_scope` to prefix some variables with discriminator so that we can retrieve and reuse them.

So, let's define the function which will return the binary output of the discriminator as well as the logit values:

```
# Defining the discriminator function
def discriminator(input_imgs, reuse=False):
    # using variable_scope to reuse variables
    with tf.variable_scope('discriminator', reuse=reuse):
        # leaky relu parameter
```

```
            leaky_param_alpha = 0.2

            # defining the layers
            conv_layer_1 = tf.layers.conv2d(input_imgs, 64, 5, 2, 'same')
            leaky_relu_output = tf.maximum(leaky_param_alpha * conv_layer_1,
conv_layer_1)

            conv_layer_2 = tf.layers.conv2d(leaky_relu_output, 128, 5, 2,
'same')
            normalized_output = tf.layers.batch_normalization(conv_layer_2,
training=True)
            leay_relu_output = tf.maximum(leaky_param_alpha *
normalized_output, normalized_output)

            conv_layer_3 = tf.layers.conv2d(leay_relu_output, 256, 5, 2,
'same')
            normalized_output = tf.layers.batch_normalization(conv_layer_3,
training=True)
            leaky_relu_output = tf.maximum(leaky_param_alpha *
normalized_output, normalized_output)

            # reshaping the output for the logits to be 2D tensor
            flattened_output = tf.reshape(leaky_relu_output, (-1, 4 * 4 * 256))
            logits_layer = tf.layers.dense(flattened_output, 1)
            output = tf.sigmoid(logits_layer)

        return output, logits_layer
```

Generator

Now, it's time to implement the second part of the network that will be trying to replicate the original input images using the latent space z. We'll be using tf.variable_scope for this function as well.

So, let's define the function which will return a generated image by the generator:

```
    def generator(z_latent_space, output_channel_dim, is_train=True):

        with tf.variable_scope('generator', reuse=not is_train):
            #leaky relu parameter
            leaky_param_alpha = 0.2
            fully_connected_layer = tf.layers.dense(z_latent_space, 2*2*512)
            #reshaping the output back to 4D tensor to match the accepted
    format for convolution layer
            reshaped_output = tf.reshape(fully_connected_layer, (-1, 2, 2,
    512))
```

```
        normalized_output = tf.layers.batch_normalization(reshaped_output,
training=is_train)
        leaky_relu_output = tf.maximum(leaky_param_alpha *
normalized_output, normalized_output)
        conv_layer_1 = tf.layers.conv2d_transpose(leaky_relu_output, 256,
5, 2, 'valid')
        normalized_output = tf.layers.batch_normalization(conv_layer_1,
training=is_train)
        leaky_relu_output = tf.maximum(leaky_param_alpha *
normalized_output, normalized_output)
        conv_layer_2 = tf.layers.conv2d_transpose(leaky_relu_output, 128,
5, 2, 'same')
        normalized_output = tf.layers.batch_normalization(conv_layer_2,
training=is_train)
        leaky_relu_output = tf.maximum(leaky_param_alpha *
normalized_output, normalized_output)
        logits_layer = tf.layers.conv2d_transpose(leaky_relu_output,
output_channel_dim, 5, 2, 'same')
        output = tf.tanh(logits_layer)
        return output
```

Model losses

Now comes the tricky part, which we covered in the previous chapter, which is to calculate the losses of the discriminator and the generator.

So, let's define this function, which will make use of the `generator` and `discriminator` functions that were defined previously:

```
# Define the error for the discriminator and generator
def model_losses(input_actual, input_latent_z, out_channel_dim):
    # building the generator part
    gen_model = generator(input_latent_z, out_channel_dim)
    disc_model_true, disc_logits_true = discriminator(input_actual)
    disc_model_fake, disc_logits_fake = discriminator(gen_model,
reuse=True)

    disc_loss_true = tf.reduce_mean(
        tf.nn.sigmoid_cross_entropy_with_logits(logits=disc_logits_true,
labels=tf.ones_like(disc_model_true)))

    disc_loss_fake =
tf.reduce_mean(tf.nn.sigmoid_cross_entropy_with_logits(
        logits=disc_logits_fake, labels=tf.zeros_like(disc_model_fake)))

    gen_loss = tf.reduce_mean(tf.nn.sigmoid_cross_entropy_with_logits(
```

```
        logits=disc_logits_fake, labels=tf.ones_like(disc_model_fake)))

    disc_loss = disc_loss_true + disc_loss_fake

    return disc_loss, gen_loss
```

Model optimizer

Finally, before training our model, we need to implement the optimization criteria for this task. We will use the naming conventions that we used previously to retrieve the trainable parameters for the discriminator and the generator and train them:

```
# specifying the optimization criteria
def model_optimizer(disc_loss, gen_loss, learning_rate, beta1):
    trainable_vars = tf.trainable_variables()
    disc_vars = [var for var in trainable_vars if
var.name.startswith('discriminator')]
    gen_vars = [var for var in trainable_vars if
var.name.startswith('generator')]

    disc_train_opt = tf.train.AdamOptimizer(
        learning_rate, beta1=beta1).minimize(disc_loss, var_list=disc_vars)

    update_operations = tf.get_collection(tf.GraphKeys.UPDATE_OPS)
    gen_updates = [opt for opt in update_operations if
opt.name.startswith('generator')]

    with tf.control_dependencies(gen_updates):
        gen_train_opt = tf.train.AdamOptimizer(
            learning_rate, beta1).minimize(gen_loss, var_list=gen_vars)

    return disc_train_opt, gen_train_opt
```

Training the model

Now, it's time to train the model and see how the generator will be able to fool, to some extent, the discriminator, by generating images very close to the original CelebA dataset.

But first, let's define a helper function that will display some generated images by the generator:

```
# define a function to visualize some generated images from the generator
def show_generator_output(sess, num_images, input_latent_z,
output_channel_dim, img_mode):
    cmap = None if img_mode == 'RGB' else 'gray'
```

```
    latent_space_z_dim = input_latent_z.get_shape().as_list()[-1]
    examples_z = np.random.uniform(-1, 1, size=[num_images,
latent_space_z_dim])

    examples = sess.run(
        generator(input_latent_z, output_channel_dim, False),
        feed_dict={input_latent_z: examples_z})

    images_grid = utils.images_square_grid(examples, img_mode)
    pyplot.imshow(images_grid, cmap=cmap)
    pyplot.show()
```

Then, we will use the helper functions that we have defined before to build the model inputs, loss, and optimization criteria. We stack them together and start training our model based on the CelebA dataset:

```
def model_train(num_epocs, train_batch_size, z_dim, learning_rate, beta1,
get_batches, input_data_shape, data_img_mode):
    _, image_width, image_height, image_channels = input_data_shape

    actual_input, z_input, leaningRate = inputs(
        image_width, image_height, image_channels, z_dim)

    disc_loss, gen_loss = model_losses(actual_input, z_input,
image_channels)

    disc_opt, gen_opt = model_optimizer(disc_loss, gen_loss, learning_rate,
beta1)

    steps = 0
    print_every = 50
    show_every = 100
    model_loss = []
    num_images = 25

    with tf.Session() as sess:

        # initializing all the variables
        sess.run(tf.global_variables_initializer())

        for epoch_i in range(num_epocs):
            for batch_images in get_batches(train_batch_size):

                steps += 1
                batch_images *= 2.0
                z_sample = np.random.uniform(-1, 1, (train_batch_size,
z_dim))
```

```
                    _ = sess.run(disc_opt, feed_dict={
                        actual_input: batch_images, z_input: z_sample,
    leaningRate: learning_rate})
                        _ = sess.run(gen_opt, feed_dict={
                        z_input: z_sample, leaningRate: learning_rate})

                    if steps % print_every == 0:
                        train_loss_disc = disc_loss.eval({z_input: z_sample,
    actual_input: batch_images})
                        train_loss_gen = gen_loss.eval({z_input: z_sample})

                        print("Epoch {}/{}...".format(epoch_i + 1, num_epocs),
                                "Discriminator Loss:
    {:.4f}...".format(train_loss_disc),
                                "Generator Loss: {:.4f}".format(train_loss_gen))
                        model_loss.append((train_loss_disc, train_loss_gen))

                    if steps % show_every == 0:
                        show_generator_output(sess, num_images, z_input,
    image_channels, data_img_mode)
```

Kick off the training process, which might take some time depending on your host machine specs:

```
# Training the model on CelebA dataset
train_batch_size = 64
z_dim = 100
learning_rate = 0.002
beta1 = 0.5

num_epochs = 1

celeba_dataset = utils.Dataset('celeba', glob(os.path.join(data_dir,
'img_align_celeba/*.jpg')))
with tf.Graph().as_default():
    model_train(num_epochs, train_batch_size, z_dim, learning_rate, beta1,
celeba_dataset.get_batches,
                celeba_dataset.shape, celeba_dataset.image_mode)
```

Output:

```
    Epoch 1/1... Discriminator Loss: 0.9118... Generator Loss: 12.2238
    Epoch 1/1... Discriminator Loss: 0.6119... Generator Loss: 3.2168
    Epoch 1/1... Discriminator Loss: 0.5383... Generator Loss: 2.8054
    Epoch 1/1... Discriminator Loss: 1.4381... Generator Loss: 0.4672
    Epoch 1/1... Discriminator Loss: 0.7815... Generator Loss: 14.8220
    Epoch 1/1... Discriminator Loss: 0.6435... Generator Loss: 9.2591
    Epoch 1/1... Discriminator Loss: 1.5661... Generator Loss: 10.4747
```

```
Epoch 1/1... Discriminator Loss: 1.5407... Generator Loss: 0.5811
Epoch 1/1... Discriminator Loss: 0.6470... Generator Loss: 2.9002
Epoch 1/1... Discriminator Loss: 0.5671... Generator Loss: 2.0700
```

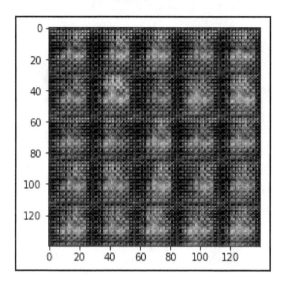

Figure 3: Sample generated output at this point of training

```
Epoch 1/1... Discriminator Loss: 0.7950... Generator Loss: 1.5818
Epoch 1/1... Discriminator Loss: 1.2417... Generator Loss: 0.7094
Epoch 1/1... Discriminator Loss: 1.1786... Generator Loss: 1.0948
Epoch 1/1... Discriminator Loss: 1.0427... Generator Loss: 2.8878
Epoch 1/1... Discriminator Loss: 0.8409... Generator Loss: 2.6785
Epoch 1/1... Discriminator Loss: 0.8557... Generator Loss: 1.7706
Epoch 1/1... Discriminator Loss: 0.8241... Generator Loss: 1.2898
Epoch 1/1... Discriminator Loss: 0.8590... Generator Loss: 1.8217
Epoch 1/1... Discriminator Loss: 1.1694... Generator Loss: 0.8490
Epoch 1/1... Discriminator Loss: 0.9984... Generator Loss: 1.0042
```

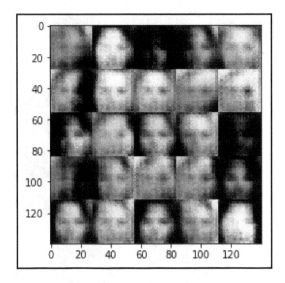

Figure 4: Sample generated output at this point of training

After some time of training, you should get something like this:

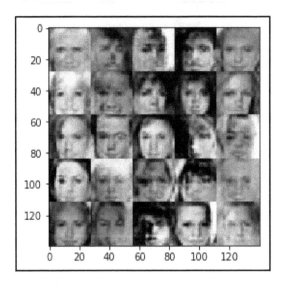

Figure 5: Sample generated output at this point of training

Semi-supervised learning with Generative Adversarial Networks (GANs)

With that in mind, semi-supervised learning is a technique in which both labeled and unlabeled data is used to train a classifier.

This type of classifier takes a tiny portion of labeled data and a much larger amount of unlabeled data (from the same domain). The goal is to combine these sources of data to train a **Deep Convolution Neural Network (DCNN)** to learn an inferred function capable of mapping a new datapoint to its desirable outcome.

In this frontier, we present a GAN model to classify street view house numbers using a very small labeled training set. In fact, the model uses roughly 1.3% of the original SVHN training labels i.e. 1000 (one thousand) labeled examples. We use some of the techniques described in the paper *Improved Techniques for Training GANs from OpenAI* (`https://arxiv.org/abs/1606.03498`).

Intuition

When building a GAN for generating images, we trained both the generator and the discriminator at the same time. After training, we can discard the discriminator because we only used it for training the generator.

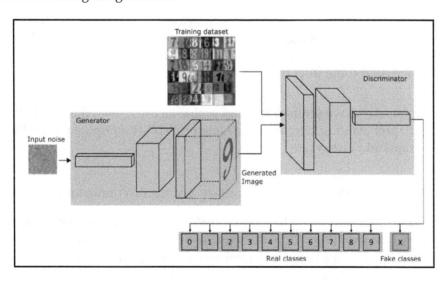

Figure 6: Semi-supervised learning GAN architecture for an 11 class classification problem

In semi-supervised learning, we need to transform the discriminator into a multi-class classifier. This new model has to be able to generalize well on the test set, even though we do not have many labeled examples for training. Additionally, this time, by the end of training, we can actually throw away the generator. Note that the roles changed. Now, the generator is only used for helping the discriminator during training. Putting it differently, the generator acts as a different source of information from which the discriminator gets raw, unlabeled training data. As we will see, this unlabelled data is key to improving the discriminator's performance. Also, for a regular image generation GAN, the discriminator has only one role. Compute the probability of whether its inputs are real or not—let's call it the GAN problem.

However, to turn the discriminator into a semi-supervised classifier, besides the GAN problem, the discriminator also has to learn the probabilities of each of the original dataset classes. In other words, for each input image, the discriminator has to learn the probabilities of it being a one, two, three, and so on.

Recall that for an image generation GAN discriminator, we have a single sigmoid unit output. This value represents the probability of an input image being real (value close to 1), or fake (value near 0). In other words, from the discriminator's point of view, values close to 1 mean that the samples are likely to come from the training set. Likewise, value near 0 mean a higher chance that the samples come from the generator network. By using this probability, the discriminator is able to send a signal back to the generator. This signal allows the generator to adapt its parameters during training, making it possible to improve its capabilities of creating realistic images.

We have to convert the discriminator (from the previous GAN) into an 11 class classifier. To do that, we can turn its sigmoid output into a softmax with 11 class outputs, the first 10 for the individual class probabilities of the SVHN dataset (zero to nine), and the 11th class for all the fake images that come from the generator.

 Note that if we set the 11th class probability to 0, then the sum of the first 10 probabilities represents the same probability computed using the sigmoid function.

Finally, we need to set up the losses in such a way that the discriminator can do both:

- Help the generator learn to produce realistic images. To do that, we have to instruct the discriminator to distinguish between real and fake samples.
- Use the generator's images, along with the labeled and unlabeled training data, to help classify the dataset.

To summarize, the discriminator has three different sources of training data:

- Real images with labels. These are image label pairs like in any regular supervised classification problem.
- Real images without labels. For those, the classifier only learns that these images are real.
- Images from the generator. To use these ones, the discriminator learns to classify as fake.

The combination of these different sources of data will make the classifier able to learn from a broader perspective. That, in turn, allows the model to perform inference much more precisely than it would be if only using the 1,000 labeled examples for training.

Data analysis and preprocessing

For this task, we will be using SVHN dataset, which is an abbreviation for Street View House Numbers by Stanford (http://ufldl.stanford.edu/housenumbers/). So, let's start the implementation by importing the required packages for this implementation:

```
# Lets start by loading the necessary libraries
%matplotlib inline

import pickle as pkl
import time
import matplotlib.pyplot as plt
import numpy as np
from scipy.io import loadmat
import tensorflow as tf
import os
```

Next up, we are going to define a helper class to download the SVHN dataset (remember that you need to manually create the input_data_dir first):

```
from urllib.request import urlretrieve
from os.path import isfile, isdir
from tqdm import tqdm
input_data_dir = 'input/'

input_data_dir = 'input/'

if not isdir(input_data_dir):
    raise Exception("Data directory doesn't exist!")

class DLProgress(tqdm):
```

```
        last_block = 0

    def hook(self, block_num=1, block_size=1, total_size=None):
        self.total = total_size
        self.update((block_num - self.last_block) * block_size)
        self.last_block = block_num

if not isfile(input_data_dir + "train_32x32.mat"):
    with DLProgress(unit='B', unit_scale=True, miniters=1, desc='SVHN
Training Set') as pbar:
        urlretrieve(
            'http://ufldl.stanford.edu/housenumbers/train_32x32.mat',
            input_data_dir + 'train_32x32.mat',
            pbar.hook)

if not isfile(input_data_dir + "test_32x32.mat"):
    with DLProgress(unit='B', unit_scale=True, miniters=1, desc='SVHN
Training Set') as pbar:
        urlretrieve(
            'http://ufldl.stanford.edu/housenumbers/test_32x32.mat',
            input_data_dir + 'test_32x32.mat',
            pbar.hook)


train_data = loadmat(input_data_dir + 'train_32x32.mat')
test_data = loadmat(input_data_dir + 'test_32x32.mat')

Output:

trainset shape: (32, 32, 3, 73257)
testset shape: (32, 32, 3, 26032)
```

Let's get a sense of what these images look like:

```
indices = np.random.randint(0, train_data['X'].shape[3], size=36)
fig, axes = plt.subplots(6, 6, sharex=True, sharey=True, figsize=(5,5),)
for ii, ax in zip(indices, axes.flatten()):
    ax.imshow(train_data['X'][:,:,:,ii], aspect='equal')
    ax.xaxis.set_visible(False)
    ax.yaxis.set_visible(False)
plt.subplots_adjust(wspace=0, hspace=0)
```

Output:

Figure 7: Sample images from the SVHN dataset.

Next up, we need to scale our images to be between -1 and 1, and this will be necessary since we are going to use the `tanh()` function, which will squash the output values of the generator:

```
# Scaling the input images
def scale_images(image, feature_range=(-1, 1)):
    # scale image to (0, 1)
    image = ((image - image.min()) / (255 - image.min()))

    # scale the image to feature range
    min, max = feature_range
    image = image * (max - min) + min
    return image

class Dataset:
    def __init__(self, train_set, test_set, validation_frac=0.5,
shuffle_data=True, scale_func=None):
        split_ind = int(len(test_set['y']) * (1 - validation_frac))
        self.test_input, self.valid_input = test_set['X'][:, :, :,
:split_ind], test_set['X'][:, :, :, split_ind:]
        self.test_target, self.valid_target = test_set['y'][:split_ind],
test_set['y'][split_ind:]
```

```
        self.train_input, self.train_target = train_set['X'],
train_set['y']

        # The street house number dataset comes with lots of labels,
        # but because we are going to do semi-supervised learning we are
going to assume that we don't have all labels
        # like, assume that we have only 1000
        self.label_mask = np.zeros_like(self.train_target)
        self.label_mask[0:1000] = 1

        self.train_input = np.rollaxis(self.train_input, 3)
        self.valid_input = np.rollaxis(self.valid_input, 3)
        self.test_input = np.rollaxis(self.test_input, 3)

        if scale_func is None:
            self.scaler = scale_images
        else:
            self.scaler = scale_func
        self.train_input = self.scaler(self.train_input)
        self.valid_input = self.scaler(self.valid_input)
        self.test_input = self.scaler(self.test_input)
        self.shuffle = shuffle_data

    def batches(self, batch_size, which_set="train"):
        input_name = which_set + "_input"
        target_name = which_set + "_target"

        num_samples = len(getattr(dataset, target_name))
        if self.shuffle:
            indices = np.arange(num_samples)
            np.random.shuffle(indices)
            setattr(dataset, input_name, getattr(dataset,
input_name)[indices])
            setattr(dataset, target_name, getattr(dataset,
target_name)[indices])
            if which_set == "train":
                dataset.label_mask = dataset.label_mask[indices]

        dataset_input = getattr(dataset, input_name)
        dataset_target = getattr(dataset, target_name)

        for jj in range(0, num_samples, batch_size):
            input_vals = dataset_input[jj:jj + batch_size]
            target_vals = dataset_target[jj:jj + batch_size]

            if which_set == "train":
                # including the label mask in case of training
                # to pretend that we don't have all the labels
```

```
                    yield input_vals, target_vals, self.label_mask[jj:jj +
batch_size]
            else:
                yield input_vals, target_vals
```

Building the model

In this section, we will build all the bits and pieces that are necessary for our test, so let's start off by defining the inputs that will be used to feed data to the computational graph.

Model inputs

First off, we are going to define the model inputs function, which will create the model input placeholders to be used for feeding data to the computational model:

```
# defining the model inputs
def inputs(actual_dim, z_dim):
    inputs_actual = tf.placeholder(tf.float32, (None, *actual_dim),
name='input_actual')
    inputs_latent_z = tf.placeholder(tf.float32, (None, z_dim),
name='input_latent_z')

    target = tf.placeholder(tf.int32, (None), name='target')
    label_mask = tf.placeholder(tf.int32, (None), name='label_mask')

    return inputs_actual, inputs_latent_z, target, label_mask
```

Generator

In this section, we are going to implement the first core part of the GAN network. The architecture and implementation of this part will follow the original DCGAN paper:

```
def generator(latent_z, output_image_dim, reuse_vars=False,
leaky_alpha=0.2, is_training=True, size_mult=128):
    with tf.variable_scope('generator', reuse=reuse_vars):
        # define a fully connected layer
        fully_conntected_1 = tf.layers.dense(latent_z, 4 * 4 * size_mult *
4)

        # Reshape it from 2D tensor to 4D tensor to be fed to the
convolution neural network
        reshaped_out_1 = tf.reshape(fully_conntected_1, (-1, 4, 4,
size_mult * 4))
```

```
        batch_normalization_1 =
tf.layers.batch_normalization(reshaped_out_1, training=is_training)
        leaky_output_1 = tf.maximum(leaky_alpha * batch_normalization_1,
batch_normalization_1)

        conv_layer_1 = tf.layers.conv2d_transpose(leaky_output_1, size_mult
* 2, 5, strides=2, padding='same')
        batch_normalization_2 = tf.layers.batch_normalization(conv_layer_1,
training=is_training)
        leaky_output_2 = tf.maximum(leaky_alpha * batch_normalization_2,
batch_normalization_2)

        conv_layer_2 = tf.layers.conv2d_transpose(leaky_output_2,
size_mult, 5, strides=2, padding='same')
        batch_normalization_3 = tf.layers.batch_normalization(conv_layer_2,
training=is_training)
        leaky_output_3 = tf.maximum(leaky_alpha * batch_normalization_3,
batch_normalization_3)

        # defining the output layer
        logits_layer = tf.layers.conv2d_transpose(leaky_output_3,
output_image_dim, 5, strides=2, padding='same')

        output = tf.tanh(logits_layer)

        return output
```

Discriminator

Now, it's time to build the second core piece of the GAN network, which is the discriminator. In previous implementations, we said that the discriminator will produce a binary output that represents whether the input image is from the real dataset (1) or it's generated by the generator (0). The scenario is different here, so the discriminator will now be a multi-class classifier.

Now, let's go ahead and build up the discriminator part of the architecture:

```
# Defining the discriminator part of the network
def discriminator(input_x, reuse_vars=False, leaky_alpha=0.2,
drop_out_rate=0., num_classes=10, size_mult=64):
    with tf.variable_scope('discriminator', reuse=reuse_vars):

        # defining a dropout layer
        drop_out_output = tf.layers.dropout(input_x, rate=drop_out_rate /
2.5)
```

```
        # Defining the input layer for the discriminator which is 32x32x3
        conv_layer_3 = tf.layers.conv2d(input_x, size_mult, 3, strides=2,
padding='same')
        leaky_output_4 = tf.maximum(leaky_alpha * conv_layer_3,
conv_layer_3)
        leaky_output_4 = tf.layers.dropout(leaky_output_4,
rate=drop_out_rate)

        conv_layer_4 = tf.layers.conv2d(leaky_output_4, size_mult, 3,
strides=2, padding='same')
        batch_normalization_4 = tf.layers.batch_normalization(conv_layer_4,
training=True)
        leaky_output_5 = tf.maximum(leaky_alpha * batch_normalization_4,
batch_normalization_4)

        conv_layer_5 = tf.layers.conv2d(leaky_output_5, size_mult, 3,
strides=2, padding='same')
        batch_normalization_5 = tf.layers.batch_normalization(conv_layer_5,
training=True)
        leaky_output_6 = tf.maximum(leaky_alpha * batch_normalization_5,
batch_normalization_5)
        leaky_output_6 = tf.layers.dropout(leaky_output_6,
rate=drop_out_rate)

        conv_layer_6 = tf.layers.conv2d(leaky_output_6, 2 * size_mult, 3,
strides=1, padding='same')
        batch_normalization_6 = tf.layers.batch_normalization(conv_layer_6,
training=True)
        leaky_output_7 = tf.maximum(leaky_alpha * batch_normalization_6,
batch_normalization_6)

        conv_layer_7 = tf.layers.conv2d(leaky_output_7, 2 * size_mult, 3,
strides=1, padding='same')
        batch_normalization_7 = tf.layers.batch_normalization(conv_layer_7,
training=True)
        leaky_output_8 = tf.maximum(leaky_alpha * batch_normalization_7,
batch_normalization_7)

        conv_layer_8 = tf.layers.conv2d(leaky_output_8, 2 * size_mult, 3,
strides=2, padding='same')
        batch_normalization_8 = tf.layers.batch_normalization(conv_layer_8,
training=True)
        leaky_output_9 = tf.maximum(leaky_alpha * batch_normalization_8,
batch_normalization_8)
        leaky_output_9 = tf.layers.dropout(leaky_output_9,
rate=drop_out_rate)

        conv_layer_9 = tf.layers.conv2d(leaky_output_9, 2 * size_mult, 3,
```

```
strides=1, padding='valid')

        leaky_output_10 = tf.maximum(leaky_alpha * conv_layer_9,
conv_layer_9)

    ...
```

Instead of applying a fully connected layer at the end, we are going to perform so-called **global average pooling** (**GAP**), which takes the average over the spatial dimensions of a feature vector; this will produce a squashed tensor to only a single value:

```
    ...

# Flatten it by global average pooling
leaky_output_features = tf.reduce_mean(leaky_output_10, (1, 2))

    ...
```

For example, suppose that after a stack of convolutions, we get an output tensor of shape:

```
[BATCH_SIZE, 8, 8, NUM_CHANNELS]
```

To apply global average pooling, we calculate the average value on the [8x8] tensor slice. This operation will result in a tensor which is the following shape:

```
[BATCH_SIZE, 1, 1, NUM_CHANNELS]
```

That can be reshaped to:

```
[BATCH_SIZE, NUM_CHANNELS].
```

After applying the global average pooling, we add a fully connected layer that will output the final logits. These have the shape of:

```
[BATCH_SIZE, NUM_CLASSES]
```

which will represent the scores for each class. To get these scores for probability, we are going to use the `softmax` activation function:

```
    ...
# Get the probability that the input is real rather than fake
softmax_output = tf.nn.softmax(classes_logits)s
    ...
```

And finally the discriminator function will look like this,

```
# Defining the discriminator part of the network
def discriminator(input_x, reuse_vars=False, leaky_alpha=0.2,
drop_out_rate=0., num_classes=10, size_mult=64):
    with tf.variable_scope('discriminator', reuse=reuse_vars):

        # defining a dropout layer
        drop_out_output = tf.layers.dropout(input_x, rate=drop_out_rate /
2.5)

        # Defining the input layer for the discrminator which is 32x32x3
        conv_layer_3 = tf.layers.conv2d(input_x, size_mult, 3, strides=2,
padding='same')
        leaky_output_4 = tf.maximum(leaky_alpha * conv_layer_3,
conv_layer_3)
        leaky_output_4 = tf.layers.dropout(leaky_output_4,
rate=drop_out_rate)

        conv_layer_4 = tf.layers.conv2d(leaky_output_4, size_mult, 3,
strides=2, padding='same')
        batch_normalization_4 = tf.layers.batch_normalization(conv_layer_4,
training=True)
        leaky_output_5 = tf.maximum(leaky_alpha * batch_normalization_4,
batch_normalization_4)

        conv_layer_5 = tf.layers.conv2d(leaky_output_5, size_mult, 3,
strides=2, padding='same')
        batch_normalization_5 = tf.layers.batch_normalization(conv_layer_5,
training=True)
        leaky_output_6 = tf.maximum(leaky_alpha * batch_normalization_5,
batch_normalization_5)
        leaky_output_6 = tf.layers.dropout(leaky_output_6,
rate=drop_out_rate)

        conv_layer_6 = tf.layers.conv2d(leaky_output_6, 2 * size_mult, 3,
strides=1, padding='same')
        batch_normalization_6 = tf.layers.batch_normalization(conv_layer_6,
training=True)
        leaky_output_7 = tf.maximum(leaky_alpha * batch_normalization_6,
batch_normalization_6)

        conv_layer_7 = tf.layers.conv2d(leaky_output_7, 2 * size_mult, 3,
strides=1, padding='same')
        batch_normalization_7 = tf.layers.batch_normalization(conv_layer_7,
training=True)
        leaky_output_8 = tf.maximum(leaky_alpha * batch_normalization_7,
batch_normalization_7)
```

```
        conv_layer_8 = tf.layers.conv2d(leaky_output_8, 2 * size_mult, 3,
strides=2, padding='same')
        batch_normalization_8 = tf.layers.batch_normalization(conv_layer_8,
training=True)
        leaky_output_9 = tf.maximum(leaky_alpha * batch_normalization_8,
batch_normalization_8)
        leaky_output_9 = tf.layers.dropout(leaky_output_9,
rate=drop_out_rate)

        conv_layer_9 = tf.layers.conv2d(leaky_output_9, 2 * size_mult, 3,
strides=1, padding='valid')

        leaky_output_10 = tf.maximum(leaky_alpha * conv_layer_9,
conv_layer_9)

        # Flatten it by global average pooling
        leaky_output_features = tf.reduce_mean(leaky_output_10, (1, 2))

        # Set class_logits to be the inputs to a softmax distribution over
the different classes
        classes_logits = tf.layers.dense(leaky_output_features, num_classes
+ extra_class)

        if extra_class:
            actual_class_logits, fake_class_logits =
tf.split(classes_logits, [num_classes, 1], 1)
            assert fake_class_logits.get_shape()[1] == 1,
fake_class_logits.get_shape()
            fake_class_logits = tf.squeeze(fake_class_logits)
        else:
            actual_class_logits = classes_logits
            fake_class_logits = 0.

        max_reduced = tf.reduce_max(actual_class_logits, 1, keep_dims=True)
        stable_actual_class_logits = actual_class_logits - max_reduced

        gan_logits =
tf.log(tf.reduce_sum(tf.exp(stable_actual_class_logits), 1)) + tf.squeeze(
            max_reduced) - fake_class_logits

        softmax_output = tf.nn.softmax(classes_logits)

        return softmax_output, classes_logits, gan_logits,
leaky_output_features
```

Model losses

Now it's time to define the model losses. First off, the discriminator loss will be divided into two parts:

- One which will represent the GAN problem, which is the unsupervised loss
- The second one will compute the individual actual class probabilities, which is the supervised loss

For the discriminator's unsupervised loss, it has to discriminate between actual training images and the generated images by the generator.

As for a regular GAN, half of the time, the discriminator will get unlabeled images from the training set as an input and the other half, fake, unlabeled images from the generator.

For the second part of the discriminator loss, which is the supervised loss, we need to build upon the logits from the discriminator. So, we will use the softmax cross entropy since it's a multi classification problem.

As mentioned in the *Enhanced Techniques for Training GANs* paper, we should use feature matching for the generator loss. As the authors describe:

> *"Feature matching is the concept of penalizing the mean absolute error between the average value of some set of features on the training data and the average values of that set of features on the generated samples. To do that, we take some set of statistics (the moments) from two different sources and force them to be similar. First, we take the average of the features extracted from the discriminator when a real training minibatch is being processed. Second, we compute the moments in the same way, but now for when a minibatch composed of fake images that come from the generator was being analyzed by the discriminator. Finally, with these two sets of moments, the generator loss is the mean absolute difference between them. In other words, as the paper emphasizes: We train the generator to match the expected values of the features on an intermediate layer of the discriminator."*

And finally, the model loss function will look like this:

```
def model_losses(input_actual, input_latent_z, output_dim, target,
num_classes, label_mask, leaky_alpha=0.2,
                drop_out_rate=0.):

        # These numbers multiply the size of each layer of the generator
    and the discriminator,
```

```
        # respectively. You can reduce them to run your code faster for
debugging purposes.
        gen_size_mult = 32
        disc_size_mult = 64

        # Here we run the generator and the discriminator
        gen_model = generator(input_latent_z, output_dim,
leaky_alpha=leaky_alpha, size_mult=gen_size_mult)
        disc_on_data = discriminator(input_actual, leaky_alpha=leaky_alpha,
drop_out_rate=drop_out_rate,
                                    size_mult=disc_size_mult)
        disc_model_real, class_logits_on_data, gan_logits_on_data,
data_features = disc_on_data
        disc_on_samples = discriminator(gen_model, reuse_vars=True,
leaky_alpha=leaky_alpha,
                                        drop_out_rate=drop_out_rate,
size_mult=disc_size_mult)
        disc_model_fake, class_logits_on_samples, gan_logits_on_samples,
sample_features = disc_on_samples

        # Here we compute `disc_loss`, the loss for the discriminator.
        disc_loss_actual = tf.reduce_mean(
tf.nn.sigmoid_cross_entropy_with_logits(logits=gan_logits_on_data,
labels=tf.ones_like(gan_logits_on_data)))
        disc_loss_fake = tf.reduce_mean(
tf.nn.sigmoid_cross_entropy_with_logits(logits=gan_logits_on_samples,
labels=tf.zeros_like(gan_logits_on_samples)))
        target = tf.squeeze(target)
        classes_cross_entropy =
tf.nn.softmax_cross_entropy_with_logits(logits=class_logits_on_data,
labels=tf.one_hot(target,
num_classes + extra_class,
dtype=tf.float32))
        classes_cross_entropy = tf.squeeze(classes_cross_entropy)
        label_m = tf.squeeze(tf.to_float(label_mask))
        disc_loss_class = tf.reduce_sum(label_m * classes_cross_entropy) /
tf.maximum(1., tf.reduce_sum(label_m))
        disc_loss = disc_loss_class + disc_loss_actual + disc_loss_fake

        # Here we set `gen_loss` to the "feature matching" loss invented by
Tim Salimans.
        sampleMoments = tf.reduce_mean(sample_features, axis=0)
        dataMoments = tf.reduce_mean(data_features, axis=0)

        gen_loss = tf.reduce_mean(tf.abs(dataMoments - sampleMoments))

        prediction_class = tf.cast(tf.argmax(class_logits_on_data, 1),
tf.int32)
```

```
    check_prediction = tf.equal(tf.squeeze(target), prediction_class)
    correct = tf.reduce_sum(tf.to_float(check_prediction))
    masked_correct = tf.reduce_sum(label_m *
tf.to_float(check_prediction))

    return disc_loss, gen_loss, correct, masked_correct, gen_model
```

Model optimizer

Now, let's define the model optimizer, which is pretty much similar to the ones that we defined before:

```
def model_optimizer(disc_loss, gen_loss, learning_rate, beta1):

    # Get weights and biases to update. Get them separately for the
discriminator and the generator
    trainable_vars = tf.trainable_variables()
    disc_vars = [var for var in trainable_vars if
var.name.startswith('discriminator')]
    gen_vars = [var for var in trainable_vars if
var.name.startswith('generator')]
    for t in trainable_vars:
        assert t in disc_vars or t in gen_vars

    # Minimize both gen and disc costs simultaneously
    disc_train_optimizer = tf.train.AdamOptimizer(learning_rate,
beta1=beta1).minimize(disc_loss,
var_list=disc_vars)
    gen_train_optimizer = tf.train.AdamOptimizer(learning_rate,
beta1=beta1).minimize(gen_loss, var_list=gen_vars)
    shrink_learning_rate = tf.assign(learning_rate, learning_rate *
0.9)

    return disc_train_optimizer, gen_train_optimizer,
shrink_learning_rate
```

Model training

Finally, let's go ahead and kick off the training process after putting it all together:

```python
class GAN:
        def __init__(self, real_size, z_size, learning_rate,
num_classes=10, alpha=0.2, beta1=0.5):
            tf.reset_default_graph()

            self.learning_rate = tf.Variable(learning_rate,
trainable=False)
            model_inputs = inputs(real_size, z_size)
            self.input_actual, self.input_latent_z, self.target,
self.label_mask = model_inputs
            self.drop_out_rate = tf.placeholder_with_default(.5, (),
"drop_out_rate")

            losses_results = model_losses(self.input_actual,
self.input_latent_z,
                                            real_size[2], self.target,
num_classes,
                                            label_mask=self.label_mask,
                                            leaky_alpha=0.2,
                                            drop_out_rate=self.drop_out_rate)
            self.disc_loss, self.gen_loss, self.correct,
self.masked_correct, self.samples = losses_results

            self.disc_opt, self.gen_opt, self.shrink_learning_rate =
model_optimizer(self.disc_loss, self.gen_loss,
self.learning_rate, beta1)

def view_generated_samples(epoch, samples, nrows, ncols, figsize=(5, 5)):
        fig, axes = plt.subplots(figsize=figsize, nrows=nrows, ncols=ncols,
                                sharey=True, sharex=True)
        for ax, img in zip(axes.flatten(), samples[epoch]):
            ax.axis('off')
            img = ((img - img.min()) * 255 / (img.max() -
img.min()))).astype(np.uint8)
            ax.set_adjustable('box-forced')
            im = ax.imshow(img)

        plt.subplots_adjust(wspace=0, hspace=0)
        return fig, axes

def train(net, dataset, epochs, batch_size, figsize=(5, 5)):

        saver = tf.train.Saver()
        sample_z = np.random.normal(0, 1, size=(50, latent_space_z_size))
```

```
        samples, train_accuracies, test_accuracies = [], [], []
        steps = 0

        with tf.Session() as sess:
            sess.run(tf.global_variables_initializer())
            for e in range(epochs):
                print("Epoch", e)

                num_samples = 0
                num_correct_samples = 0
                for x, y, label_mask in dataset.batches(batch_size):
                    assert 'int' in str(y.dtype)
                    steps += 1
                    num_samples += label_mask.sum()

                    # Sample random noise for G
                    batch_z = np.random.normal(0, 1, size=(batch_size,
latent_space_z_size))

                    _, _, correct = sess.run([net.disc_opt, net.gen_opt,
net.masked_correct],
                                                feed_dict={net.input_actual:
x, net.input_latent_z: batch_z,
                                                        net.target: y,
net.label_mask: label_mask})
                    num_correct_samples += correct

                sess.run([net.shrink_learning_rate])

                training_accuracy = num_correct_samples /
float(num_samples)

                print("\t\tClassifier train accuracy: ", training_accuracy)

                num_samples = 0
                num_correct_samples = 0

                for x, y in dataset.batches(batch_size, which_set="test"):
                    assert 'int' in str(y.dtype)
                    num_samples += x.shape[0]

                    correct, = sess.run([net.correct],
feed_dict={net.input_real: x,
                                                        net.y: y,
net.drop_rate: 0.})
                    num_correct_samples += correct

                testing_accuracy = num_correct_samples / float(num_samples)
```

```
                print("\t\tClassifier test accuracy", testing_accuracy)

                gen_samples = sess.run(
                    net.samples,
                    feed_dict={net.input_latent_z: sample_z})
                samples.append(gen_samples)
                _ = view_generated_samples(-1, samples, 5, 10,
        figsize=figsize)
                plt.show()

                # Save history of accuracies to view after training
                train_accuracies.append(training_accuracy)
                test_accuracies.append(testing_accuracy)

            saver.save(sess, './checkpoints/generator.ckpt')

        with open('samples.pkl', 'wb') as f:
            pkl.dump(samples, f)

        return train_accuracies, test_accuracies, samples
```

Don't forget to create a directory called checkpoints:

```
real_size = (32,32,3)
latent_space_z_size = 100
learning_rate = 0.0003

net = GAN(real_size, latent_space_z_size, learning_rate)

dataset = Dataset(train_data, test_data)

train_batch_size = 128
num_epochs = 25
train_accuracies, test_accuracies, samples = train(net,
                                                   dataset,
                                                   num_epochs,
                                                   train_batch_size,
                                                   figsize=(10,5))
```

Finally, at `Epoch 24`, you should get something close to this:

```
Epoch 24
                Classifier train accuracy:  0.937
                Classifier test accuracy 0.67401659496
                Step time:  0.03694915771484375
                Epoch time:  26.15842580795288
```

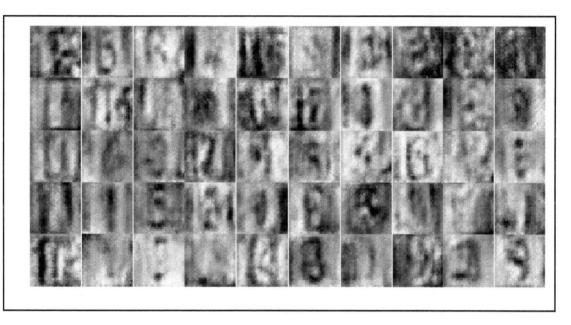

Figure 8: Sample images created by the generator network using the feature matching loss

```
fig, ax = plt.subplots()
plt.plot(train_accuracies, label='Train', alpha=0.5)
plt.plot(test_accuracies, label='Test', alpha=0.5)
plt.title("Accuracy")
plt.legend()
```

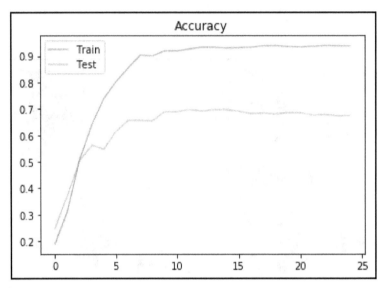

Figure 9: Train versus Test accuracy over the training process

Although feature matching loss performs well on the task of semi-supervised learning, the images produced by the generator are not as good as the ones created in the previous chapter. But this implementation was mainly introduced to demonstrate how we can use GANs for semi-supervised learning setups.

Summary

Finally, many researchers consider unsupervised learning as the missing link in general AI systems. To overcome these obstacles, attempts to solve established problems using less labeled data is key. In this scenario, GANs pose a real alternative for learning complicated tasks with less labeled samples. Yet, the performance gap between supervised and semi-supervised learning is still far from being equal. We can certainly expect this gap to become shorter as new approaches come into play.

Implementing Fish Recognition

The following is the entire piece of code of the *Fish Recognition* section covered in Chapter 1, *Data Science – Bird's-Eye View*.

Code for fish recognition

After explaining the main building blocks for our fish recognition example, we are ready to see all the code pieces connected together and see how we managed to build such a complex system with just a few lines of code:

```python
#Loading the required libraries along with the deep learning platform Keras
with TensorFlow as backend
import numpy as np
np.random.seed(2017)
import os
import glob
import cv2
import pandas as pd
import time
import warnings
from sklearn.cross_validation import KFold
from keras.models import Sequential
from keras.layers.core import Dense, Dropout, Flatten
from keras.layers.convolutional import Convolution2D, MaxPooling2D,
ZeroPadding2D
from keras.optimizers import SGD
from keras.callbacks import EarlyStopping
from keras.utils import np_utils
from sklearn.metrics import log_loss
from keras import __version__ as keras_version
# Parameters
```

```
# ----------
# x : type
#     Description of parameter `x`.
def rezize_image(img_path):
  img = cv2.imread(img_path)
  img_resized = cv2.resize(img, (32, 32), cv2.INTER_LINEAR)
  return img_resized
#Loading the training samples from their corresponding folder names, where
we have a folder for each type
def load_training_samples():
  #Variables to hold the training input and output variables
  train_input_variables = []
  train_input_variables_id = []
  train_label = []
  # Scanning all images in each folder of a fish type
  print('Start Reading Train Images')
  folders = ['ALB', 'BET', 'DOL', 'LAG', 'NoF', 'OTHER', 'SHARK', 'YFT']
  for fld in folders:
      folder_index = folders.index(fld)
      print('Load folder {} (Index: {})'.format(fld, folder_index))
      imgs_path = os.path.join('..', 'input', 'train', fld, '*.jpg')
      files = glob.glob(imgs_path)
      for file in files:
          file_base = os.path.basename(file)
          # Resize the image
          resized_img = rezize_image(file)
          # Appending the processed image to the input/output variables of
the classifier
          train_input_variables.append(resized_img)
          train_input_variables_id.append(file_base)
          train_label.append(folder_index)
  return train_input_variables, train_input_variables_id, train_label
#Loading the testing samples which will be used to testing how well the
model was trained
def load_testing_samples():
  # Scanning images from the test folder
  imgs_path = os.path.join('..', 'input', 'test_stg1', '*.jpg')
  files = sorted(glob.glob(imgs_path))
  # Variables to hold the testing samples
  testing_samples = []
  testing_samples_id = []
  #Processing the images and appending them to the array that we have
  for file in files:
      file_base = os.path.basename(file)
      # Image resizing
      resized_img = rezize_image(file)
      testing_samples.append(resized_img)
      testing_samples_id.append(file_base)
```

```python
    return testing_samples, testing_samples_id
# formatting the images to fit our model
def format_results_for_types(predictions, test_id, info):
  model_results = pd.DataFrame(predictions, columns=['ALB', 'BET', 'DOL',
'LAG', 'NoF', 'OTHER',
    'SHARK', 'YFT'])
  model_results.loc[:, 'image'] = pd.Series(test_id,
index=model_results.index)
  sub_file = 'testOutput_' + info + '.csv'
  model_results.to_csv(sub_file, index=False)
def load_normalize_training_samples():
  # Calling the load function in order to load and resize the training
samples
  training_samples, training_label, training_samples_id =
load_training_samples()
  # Converting the loaded and resized data into Numpy format
  training_samples = np.array(training_samples, dtype=np.uint8)
  training_label = np.array(training_label, dtype=np.uint8)
  # Reshaping the training samples
  training_samples = training_samples.transpose((0, 3, 1, 2))
  # Converting the training samples and training labels into float format
  training_samples = training_samples.astype('float32')
  training_samples = training_samples / 255
  training_label = np_utils.to_categorical(training_label, 8)
  return training_samples, training_label, training_samples_id
#Loading and normalizing the testing sample to fit into our model
def load_normalize_testing_samples():
  # Calling the load function in order to load and resize the testing
samples
  testing_samples, testing_samples_id = load_testing_samples()
  # Converting the loaded and resized data into Numpy format
  testing_samples = np.array(testing_samples, dtype=np.uint8)
  # Reshaping the testing samples
  testing_samples = testing_samples.transpose((0, 3, 1, 2))
  # Converting the testing samples into float format
  testing_samples = testing_samples.astype('float32')
  testing_samples = testing_samples / 255
  return testing_samples, testing_samples_id
def merge_several_folds_mean(data, num_folds):
  a = np.array(data[0])
  for i in range(1, num_folds):
      a += np.array(data[i])
  a /= num_folds
  return a.tolist()
# Create CNN model architecture
def create_cnn_model_arch():
  pool_size = 2 # we will use 2x2 pooling throughout
  conv_depth_1 = 32 # we will initially have 32 kernels per conv. layer...
```

```
conv_depth_2 = 64 # ...switching to 64 after the first pooling layer
kernel_size = 3 # we will use 3x3 kernels throughout
drop_prob = 0.5 # dropout in the FC layer with probability 0.5
hidden_size = 32 # the FC layer will have 512 neurons
num_classes = 8 # there are 8 fish types
# Conv [32] -> Conv [32] -> Pool
cnn_model = Sequential()
cnn_model.add(ZeroPadding2D((1, 1), input_shape=(3, 32, 32),
dim_ordering='th'))
cnn_model.add(Convolution2D(conv_depth_1, kernel_size, kernel_size,
activation='relu', dim_ordering='th'))
cnn_model.add(ZeroPadding2D((1, 1), dim_ordering='th')
cnn_model.add(Convolution2D(conv_depth_1, kernel_size, kernel_size,
activation='relu', dim_ordering='th'))
cnn_model.add(MaxPooling2D(pool_size=(pool_size, pool_size), strides=(2,
2), dim_ordering='th'))
# Conv [64] -> Conv [64] -> Pool
cnn_model.add(ZeroPadding2D((1, 1), dim_ordering='th'))
cnn_model.add(Convolution2D(conv_depth_2, kernel_size, kernel_size,
activation='relu', dim_ordering='th'))
cnn_model.add(ZeroPadding2D((1, 1), dim_ordering='th'))
cnn_model.add(Convolution2D(conv_depth_2, kernel_size, kernel_size,
activation='relu', dim_ordering='th'))
cnn_model.add(MaxPooling2D(pool_size=(pool_size, pool_size), strides=(2,
2), dim_ordering='th'))
# Now flatten to 1D, apply FC then ReLU (with dropout) and finally
softmax(output layer)
cnn_model.add(Flatten())
cnn_model.add(Dense(hidden_size, activation='relu'))
cnn_model.add(Dropout(drop_prob))
cnn_model.add(Dense(hidden_size, activation='relu'))
cnn_model.add(Dropout(drop_prob))
cnn_model.add(Dense(num_classes, activation='softmax'))
# initiating the stochastic gradient descent optimiser
stochastic_gradient_descent = SGD(lr=1e-2, decay=1e-6, momentum=0.9,
nesterov=True)
cnn_model.compile(optimizer=stochastic_gradient_descent,  # using the
stochastic gradient descent optimiser
                  loss='categorical_crossentropy')  # using the cross-
entropy loss function
return cnn_model
#Model using with kfold cross validation as a validation method
def create_model_with_kfold_cross_validation(nfolds=10):
batch_size = 16 # in each iteration, we consider 32 training examples at
once
num_epochs = 30 # we iterate 200 times over the entire training set
random_state = 51 # control the randomness for reproducibility of the
results on the same platform
```

```
# Loading and normalizing the training samples prior to feeding it to the
created CNN model
  training_samples, training_samples_target, training_samples_id =
load_normalize_training_samples()
  yfull_train = dict()
  # Providing Training/Testing indices to split data in the training
samples
  # which is splitting data into 10 consecutive folds with shuffling
  kf = KFold(len(train_id), n_folds=nfolds, shuffle=True,
random_state=random_state)
  fold_number = 0 # Initial value for fold number
  sum_score = 0 # overall score (will be incremented at each iteration)
  trained_models = [] # storing the modeling of each iteration over the
folds
  # Getting the training/testing samples based on the generated
training/testing indices by Kfold
  for train_index, test_index in kf:
      cnn_model = create_cnn_model_arch()
      training_samples_X = training_samples[train_index] # Getting the
training input variables
      training_samples_Y = training_samples_target[train_index] # Getting
the training output/label variable
      validation_samples_X = training_samples[test_index] # Getting the
validation input variables
      validation_samples_Y = training_samples_target[test_index] # Getting
the validation output/label variabl
      fold_number += 1
      print('Fold number {} out of {}'.format(fold_number, nfolds))
      callbacks = [
          EarlyStopping(monitor='val_loss', patience=3, verbose=0),
      ]
      # Fitting the CNN model giving the defined settings
      cnn_model.fit(training_samples_X, training_samples_Y,
batch_size=batch_size,
          nb_epoch=num_epochs,
              shuffle=True, verbose=2, validation_data=(validation_samples_X,
                validation_samples_Y),
              callbacks=callbacks)
      # measuring the generalization ability of the trained model based on
the validation set
      predictions_of_validation_samples =
          cnn_model.predict(validation_samples_X.astype('float32'),
batch_size=batch_size,
            verbose=2)
      current_model_score = log_loss(Y_valid,
predictions_of_validation_samples)
      print('Current model score log_loss: ', current_model_score)
      sum_score += current_model_score*len(test_index)
```

```
        # Store valid predictions
        for i in range(len(test_index)):
            yfull_train[test_index[i]] = predictions_of_validation_samples[i]
        # Store the trained model
        trained_models.append(cnn_model)
    # incrementing the sum_score value by the current model calculated score
    overall_score = sum_score/len(training_samples)
    print("Log_loss train independent avg: ", overall_score)
    #Reporting the model loss at this stage
    overall_settings_output_string = 'loss_' + str(overall_score) + '_folds_'
+ str(nfolds) + '_ep_' + str(num_epochs)
    return overall_settings_output_string, trained_models
#Testing how well the model is trained
def
test_generality_crossValidation_over_test_set(overall_settings_output_strin
g, cnn_models):
    batch_size = 16 # in each iteration, we consider 32 training examples at
once
    fold_number = 0 # fold iterator
    number_of_folds = len(cnn_models) # Creating number of folds based on the
value used in the training step
    yfull_test = [] # variable to hold overall predictions for the test set
    #executing the actual cross validation test process over the test set
    for j in range(number_of_folds):
        model = cnn_models[j]
        fold_number += 1
        print('Fold number {} out of {}'.format(fold_number,
number_of_folds))
        #Loading and normalizing testing samples
        testing_samples, testing_samples_id =
load_normalize_testing_samples()
        #Calling the current model over the current test fold
        test_prediction = model.predict(testing_samples,
batch_size=batch_size, verbose=2)
        yfull_test.append(test_prediction)
    test_result = merge_several_folds_mean(yfull_test, number_of_folds)
    overall_settings_output_string = 'loss_' + overall_settings_output_string
\
                + '_folds_' + str(number_of_folds)
    format_results_for_types(test_result, testing_samples_id,
overall_settings_output_string)
# Start the model training and testing
if __name__ == '__main__':
    info_string, models = create_model_with_kfold_cross_validation()
    test_generality_crossValidation_over_test_set(info_string, models)
```

Other Books You May Enjoy

If you enjoyed this book, you may be interested in these other books by Packt:

Deep Learning with Keras
Antonio Gulli, Sujit Pal

ISBN: 978-1-78712-842-2

- Optimize step-by-step functions on a large neural network using the Backpropagation Algorithm
- Fine-tune a neural network to improve the quality of results
- Use deep learning for image and audio processing
- Use Recursive Neural Tensor Networks (RNTNs) to outperform standard word embedding in special cases
- Identify problems for which Recurrent Neural Network (RNN) solutions are suitable
- Explore the process required to implement Autoencoders
- Evolve a deep neural network using reinforcement learning

Deep Learning with TensorFlow

Giancarlo Zaccone, Md. Rezaul Karim, Ahmed Menshawy

ISBN: 978-1-78646-978-6

- Learn about machine learning landscapes along with the historical development and progress of deep learning
- Learn about deep machine intelligence and GPU computing with the latest TensorFlow 1.x
- Access public datasets and utilize them using TensorFlow to load, process, and transform data
- Use TensorFlow on real-world datasets, including images, text, and more
- Learn how to evaluate the performance of your deep learning models
- Using deep learning for scalable object detection and mobile computing
- Train machines quickly to learn from data by exploring reinforcement learning techniques
- Explore active areas of deep learning research and applications

Leave a review - let other readers know what you think

Please share your thoughts on this book with others by leaving a review on the site that you bought it from. If you purchased the book from Amazon, please leave us an honest review on this book's Amazon page. This is vital so that other potential readers can see and use your unbiased opinion to make purchasing decisions, we can understand what our customers think about our products, and our authors can see your feedback on the title that they have worked with Packt to create. It will only take a few minutes of your time, but is valuable to other potential customers, our authors, and Packt. Thank you!

Index

www.ingramcontent.com/pod-product-compliance
Lightning Source LLC
Chambersburg PA
CBHW060646060326

40690CB00020B/4536